International Straits

Ana G. López Martín

International Straits

Concept, Classification and Rules of Passage

Springer

Professor Dr. Ana G. López Martín
University Complutense of Madrid
Ciudad Universitaria s/n
28040 Madrid
Spain
anagema@der.ucm.es

A Spanish language edition has been published in advance:
Ana Gemma López Martín, La navegación por los estrechos. Geostrategia y Derecho, Dykinson, Madrid, 2008

ISBN 978-3-642-12905-6 e-ISBN 978-3-642-12906-3
DOI 10.1007/978-3-642-12906-3
Springer Heidelberg Dordrecht London New York

Library of Congress Control Number: 2010932205

Cover design: WMXDesign GmbH, Heidelberg, Germany

Printed on acid-free paper

Springer is part of Springer Science+Business Media (www.springer.com)

To my parents, who gave me the life
To my children, who are my life
To Luis Ignacio, always in my mind, always in my heart

Contents

Foreword

The four 1958 Geneva Conventions on the Law of the Sea, which codified and progressively developed this sector of our legislation, were rather ephemeral despite the fact that they were constituent Conventions. In fact, the 1982 *United Nations Convention on the Law of the Sea* (UNCLOS) again undertook the same task with the same spirit 20 years later after a long drawn out global negotiation process in which all the marine areas and problems pending were analysed and discussed by the countries attending, and an apparently strengthened majority was attained, including the essential agreement between the principal naval powers and the third world countries, symbolised most grossly in the recognition of exclusive economic areas which were 200 miles wide in exchange for a significant alteration to the legal rules applicable to the international straits. From 1973 to 1982, the negotiations showed that there were a number of particular factors affecting the seas: "strait" countries, user countries, long range fishing countries, embedded countries, archipelagic countries, broad platform countries, etc.

In 1982 when the UNCLOS was adopted, it seemed to be a text with justified pretensions to be in force for a long period of time as the nine years of negotiations required for its adoption had taken into account the main problems pending agreement although not absolutely all. However, 25 years later this same opinion is no longer valid as the UNCLOS has real gaps concerning questions of a substantial nature. Firstly, there was the failure of the international seabed and ocean floor zone, then the high sea fishing regime especially with regard to transzonal highly migratory species, and the determination of the exterior limit of the continental platforms beyond 200 miles (which has given rise to claims by several coastal States), followed by the unjustified establishment of archipelagic statutes and the adoption of very long straight base lines over 200 miles by mixed States as well as others of lower intensity.

It is clear that during the codifying processes and in the posterior revision there are no eventualities, and, if the work carried out in 1958 was obsolete two decades later as a result of a multitude of new independent States, as a result of colonial emancipation, the patient, long and complex work of 1982 led to another basic structural change in international society: the disappearance of the Soviet Union and the Socialist Community of States. Thus, after 1958 a crisis arose as regards the traditional approach to the contradiction of interests between the north and south, as there was an evident majority of developing countries after 1982, or more specifically after 1990, when the crisis substantially affected the previous correlation of eastern and western forces. This means that, if the first codification ended in turmoil due to economic reasons, the second was affected by the ideological and geostrategic changes. This partially explains the current difficulties faced by the UNCLOS, and the need to adjust its provisions to the new global order.

In my opinion, this is relevant when the analysis of the current judicial regime on straits used for international navigation is addressed, as included in the UNCLOS. In 1982, the Soviet Union and its allies led the antagonism to the hegemonic pretensions of the other super power and its allies. Separated by a strong ideological differences, the Soviet Union and the United States, in their balanced super power roles, were going through a warm, almost pre nuptial idyll as regards their coincident geostrategic interests as both of them were attempting to achieve a regime which would enable the maximum mobility of their respective air and sea forces through all the straits, while preserving the security of their own straits. This interest was shared with their tributary organisations for their legitimate collective defence. Thus, it was relatively easy to reach agreements and achieve the reciprocal adjustment of interests at the III Conference with the majority of the countries of the third world in order to agree on fishing boats, that is to say, a new international legal regime in the straits in exchange for economic zones of 200 miles.

However, this exchange was not made without difficulties as several "strait" countries, that is to say, the coastal States of seas and oceans whose national security was essentially dependent on the effective control of their straits became a group which was strongly opposed to the new proposals to alter the regime established in 1958. This group included Spain. It must be borne in mind that, when the two super powers claimed new regulations regarding the freedom of passage of their warships and war planes in the straits, did not refer to *all* the straits or to *any* strait, but essentially to the handful of straits (between five and nine, according to the data provided at the time by the Office of the Geographer of the United States Department of State) with authentic global geostrategic importance; in other words, those which might place the whole of the world economy in danger in times of crisis depending on who might have their effective control due to the amount of traffic and the quality of the goods transported (for example, hydrocarbon on a large scale). In short, this entailed the capacity to guarantee or strangle world trade and the international economy at a given time. This handful of straits always included Gibraltar, closely involved in the strategic defensive axis and the security of Spain (the Balearic Islands–Gibraltar–Canary Islands Axis).

We have just pointed out the general aspirations of the naval powers in this matter in order to explain that, once a basic consensus was achieved on the two crucial questions (the binary classification of the straits and an introduction to the right of transit passage), the basic geostrategic interests were fulfilled and the road was clear to achieving a *package deal*. A small group of strait coastal countries showed their belligerence to the technical-legal aspects of the new regime announced, which was a sign that the stipulations agreed to failed to resolve some of the substantive problems of certain straits or gave rise to questions which were not fully answered or entailed contradictions with no solutions provided. Having said this, what was put forward at the time was the basic agreement on what were considered to be tiny legal nuisances which subsequent practice would solve.

The drastic changes in the position of some coastal States of straits after 1982 which previously had championed the right to innocent passage, as is the significant case of Spain, cannot be ignored. After the consolidation and normalisation of democracy, Spain chose to join its strategic future to that of the United States and NATO, which explains why it ceased to be a persistent objector to the new regime on international straits and embraced the advantages of the right to transit passage through the Strait of Gibraltar in order not to become involved in an insurmountable contradiction with the winds of History. In fact, the allies of Spain were strongly in favour of this and it was considered that the effective control of the strait guaranteed by the regime of 1958 was more nominal and formal than real as Spain lacked a dissuasive military force to combat nuclear submarines with strategic nuclear weapons. In fact, in 1982 the effective control of Gibraltar was located in the British military base on the rock and the United States military base in Rota, which meant that Spain was superfluous and could only be invited to the banquet after joining NATO. Thus, it had to take back what it had said earlier.

Besides the factors mentioned above regarding the reconsideration of the regime for navigation through international straits, there are other reasons endogenous to the evolution of the law of the sea which follow the same route. Among these reasons was the progressive extension of the territorial sea as, according to the 1958 regime with territorial seas usually limited to 3 nautical miles, a substantial number of straits understood from the geographical point of view were not the same in the legal sense as the breadth of the strait was greater than 6 miles and there were areas of high seas open to free navigation and overflight. However, the simple fact that the 12 mile territorial seas became generalised led to a 101 straits being considered as such in strict legal terms. In other words, the question of the straits and their legal regime ceased to be a minor question and became a common problem in practically all the geographical straits. This generalisation was due to the hasty conversion of the provisions regarding innocent passage to a general regime within less than two decades when this had only been applied to a limited number of straits since 1958.

More specifically, as regards the provisions of the 1982 UNCLOS, one of its outstanding aspects is the fact that it does not give a definition of the "straits used for international navigation", and, although this data was not of any special concern in the 1958 regime, it did become important in the 1982 regime for the reasons mentioned above. This silence was not a result of forgetfulness or improvisation during a codifying conference involving numerous political and legal controversies, but was rather due to the calculated ambiguity which would serve to understand the new right of transit passage; that is to say, this lack of definition was to the advantage of the great sea and air powers, and gave the coastal countries of straits new possibilities within this lack of definition. Thus, the provisions of Part III of the

UNCLOS were made to prevail over those of Part II, the expired and obsolete heir of the 1958 provisions, giving prevalence to the right of transit passage over the right of innocent passage, that is to say, giving more freedom to the countries using these straits, the main naval powers. This means that an objective component which cannot be appealed against (geography and geology have decided that a State is a coastal State of a strait) is to the advantage of the interests of the vessels with flags of convenience and other users.

At this point, mention should be made of the effective application of international responsibility as regards the users mentioned above. Undoubtedly the regime of the straits must maintain a balance between the general interests of international navigation involved in the transport of goods and those concerned with the security of the coastal countries. Logically, in 1982 the security of the coastal countries was focussed on military, geostrategic and defensive factors as a consequence of the bipolarity of the time; however, at the present time, this component is no longer the decisive one as the concept of security is perceived in broader terms and, in a monopolar world, the military argument is no longer relevant – especially as regards the coastal States belonging to alliances with the *hegemonic power* –, and those threatened by a serious risk of massive pollution due to accidents involving vessels transporting hydrocarbons. The most recent sea catastrophes recorded in coastal States were generally the result of sea and land pollution as a result of accidents involving oil tankers and not the result of military action. This gives rise to the question of international responsibility, especially in the case of the numerous vessels sailing with flags of convenience which systematically fail to comply with the required security conditions or which do not respect the stipulations regarding the separation and ordering of traffic in the straits laid down by the coastal States. In these cases, when a vessel in passage is involved in internationally illicit events, the possibility that the coastal State might initiate an action regarding international responsibility against the State of the vessel becomes merely virtual and illusory, especially when this country is underdeveloped. The only way to repair the damage is privately through the insurance contracted by the ship-owner, which may not exist or is simply illusory as there is a scarcity or total lack of adequate premiums.

Similar to this problem and apart from international responsibility there is the passage of other types of merchant vessels through the straits, such as nuclear driven vessels, those with special characteristics and those used for the transport of highly hazardous or toxic goods. In these cases, the risk inherent to the pollution of the waters and coasts of the coastal States multiplies, together with the risk from aircraft. In other words, the balance of interests must take into account the permanent risk of pollution felt by the coastal State; consequently, international legislation must provide precise regulations in this regard, especially when the provisions of the UNCLOS contain wide areas for development in this area. Notwithstanding the above, its guarantee rules are extremely generic and insufficient concerning the protection of the interests of the coastal State against certain users. As an example, there is article 19 of Part II (curiously related to innocent passage) for warships and warplanes related to the rules contained articles 38 et sequitur of Part III specifically dedicated to the right of transit passage. In conclusion, the Convention gives prevalence to the mobility of the large sea and air fleets over the security of the coastal State in the straits included in its Part III.

Both from the quantitative and qualitative point of view, what was common currency in the 1958 system, the right of innocent passage came to an end as from 1982 as the right of transit passage through the main straits, that is to say, those included in article 37, became dominant as regards the number of straits affected and their intrinsic importance in the context of international navigation, as well as with regard to the absence of sufficient guarantees to safeguard the security of their coastal States. The UNCLOS avoided the possibility to establish a precise catalogue of all the specific obligations affecting the user countries and the corresponding correlative rights of the coastal States. A hasty, superficial reading may lead the reader to feel that the regulation of the right of transit passage includes sufficient, necessary regulations, but a more conscientious and deeper analysis, as provided in this work, highlights the fact that this would be a vain, optical illusion with no grounds. The leaks detectable in the hull of the right of transit passage are so serious, although this may seem paradoxical, that they forebode the total sinking of the undeniable interests of the coastal States, not only as regards the defensive military risks, but also the wider risks to security.

This unavoidably leads us to a theoretical question which is of evident, practical interest. I do not refer to the codifying content, but to the progressive development in Part III of the UNCLOS, in relation to its possible consideration as new law in force and, therefore, its hypothetical opposition to all the coastal countries, regardless of whether they are parties or not to UNCLOS. This would be the equivalent of eliminating the right of innocent passage from the straits in Part III, including its possible application by the States which are not parties to UNCLOS. The problem can be examined from different legal perspectives: first, in relation to the possible maintenance of the strait countries as persistent objectors to the new regime for straits in article 37. The response cannot be classified as satisfactory and clarifying as many countries of this former group are currently in favour of the new regime (Spain), others which seemed to maintain the former regime have recently surrendered (Morocco), and only a minority of countries remains loyal to their original positions. The second perspective involves the process of creation and modification of the international norms, more specifically, the conduct of the coastal States affected. It appears to be clear that a legal regime of several centuries (the right of innocent passage) cannot be demolished in a brief period of 20 years (those years which have elapsed from the entry into force of UNCLOS) by the right of transit passage becoming implemented as law in force. Clearly this requires the opposition and clear conduct of the coastal States affected. As we have just seen, this is far from being systematic, constant and coherent practice by a sufficient number of objector States. However, it would be necessary to explore the possibility that it is precisely the vagueness, insufficiencies, contradictions and silences of the new regulation, which are sufficiently fertile field for the practice of certain coastal countries, even without opposing the new regime, might constitute a clear framework of reference in order to offset these deficiencies, a question of undoubted interest for countries such as Spain.

I refer to the national legislations both before and after 1982, but which remain in force in the first case and regulate particular aspects of navigation through the straits, as is the cases of Albania, China, Djibouti, Estonia, Finland, Italy, India, Montenegro, Samoa, Saint Vincent, Sweden, Uruguay, Yemen, etc., which, protected by the ambiguities of the UNCLOS and the possible coincidence of the 1958 Geneva system and the new 1982 regime, regulate very relevant aspects of the security of the coastal States with regard to the passage of warships, merchant vessels, those engaged in the transport of hazardous substances and others. Thus, in many cases State regulation of their straits involves a curious cocktail of the old codifying norms and the new norms to be progressively developed, with no apparent difficulties. The difference regarding the approaches explained above lies in the fact that a small group of strait countries have chosen persistent objection to the new regime (as in the cases of Iran, the United Arab Emirates or Venezuela), while others actively maintain particular positions with respect to new, very specific rules even though they are parties to UNCLOS. This entails a type of specific objection, manifested through legislation, which displaces the burden of opposition or non-opposition towards the States using these straits which either choose protests and formal opposition (with the risk this involves for the vessels flying their flags), or they accept the particular situations in specific straits. In other words, they manage to maintain the general terms of UNCLOS but they modify it as regards third party users as concerns specific rules and always in defence of interests of national security while observing a appreciable balance between the general interests of navigation and their own interests.

This possible approach is of undeniable interest for countries such as Spain, a coastal State of one of the five most important straits in the world from the geostrategic point of view, as this would make it possible for Spain to have systematic and specific regulation of certain modalities of navigation through the Strait of Gibraltar which will reasonably guarantee the interests of national security in this geographical scenario. Precisely, the position of Spain would be symptomatic in this regard as it is a coastal State which changed its position as a State which was a persistent pre-objector of the new regulation to that of a new convert to its excellences in a very short period of time. We all know the reasons for this transformation which involved making an agreement on defence cooperation with the United States and joining NATO. However, the strategic options and imperatives do not necessarily have to come into contradiction with other demands of national security, as happens with the obligations concerning precautions to be taken as regards over 60,000 vessels which pass through the Strait of Gibraltar every year, an obligation which cannot be renounced. In my opinion, this is a possibility which should be studied and applied although I believe it is not one of the priorities of the Spanish Foreign Ministry despite the problems occurring there due to accidents involving petrol tankers and the breakdown of nuclear submarines.

One final criticism of the UNCLOS which I should mention is the cataloguing of the international straits contained in its Parts II and III, and the lack of definition of the concept. When International Law defines and regulates certain marine areas which are the primary object of knowledge of other scientific disciplines, as is the case of bays, islands, archipelagos and continental platforms widely studied in geography and marine geology, it always adds authoritarian and arbitrary components. In fact, for International Law, an island, a bay or a continental platform are what the law says they are, and not what other sciences consider them to be. Clearly arbitrariness has certain limits, as an example, it does not make sense to consider a geographical accident to be a bay when it does not have the slightest appearance of a bay. This consideration is fully applicable to straits since, when the UNCLOS distinguishes between primary and secondary straits, and defines them incompletely or ambiguously; the 200 plus straits are infallibly included within one or other concept, with no possibility of escape and with no doubts. However, geography is richer, more varied and stubborn than the definitions of the jurists, and teaches us that there are straits with particular conditions which would not be included in the above distinction, or they are of a mixed nature or have decisive, differential factors which are not stipulated in law. Which regime is applicable to these straits without incurring arbitrariness incompatible with a healthy legal system? Again, geographical particularities (that is to say, objective particularities) call the law into question and also claim particular treatment in the legal regime.

Moreover, it is clear and interesting that the studies referring to the regime of straits in the comparative bibliography are almost all restricted to explaining the characteristics of the 1958 Convention on territorial seas and the UNCLOS, while extending their analyses to the work and the positions maintained at the respective codifying conferences, and, in the best cases, this involved analyses of some national legislations in this regard together with conclusions which, depending on their quality, we consider to be the last word on the matter. Nevertheless, in fact security and control measures concerning sea and navigation are adopted, respectively, by the International Maritime Organization (IMO) and by the International Civil Aviation Organization (ICAO), and knowledge of the internal acts of both organizations on the regime of each strait is essential in order to draft a diagnosis of the regime effectively applied to the straits. Thus, a large number of these studies provide merely formalist and nominal results regarding what is stipulated, but not with regard to the regime which is effectively operative in each strait, and this is more decisive as concerns such substantial and central matters as those concerning the security and control of international navigation, that is to say, to questions which are highly important for the security of the coastal States and are adopted on proposal by them.

This book does not contain any of these flaws. It evidently attempts to explain what happened after 1982, analysing the regime currently in force in each strait, and once and for all overcoming the bibliography obsessed by the UNCLOS and not by the results of its application in practice. The great majority of straits have been the subject of study concerning the regime effectively applicable to date, the result we have reached today through total or partial application or non-application of the UNCLOS. This responds to a dynamic and evolutionary perspective of International Law, as appears in reality and not in a photo of the 1982 Convention. The intention is to explain the law as it is and not as it should formally be.

In my opinion and in the light of these concepts, the following work of Professor López Martín contains a balanced analysis within the conventional abstract provisions and has a real application to the main straits in the context of the interests in question, once light has been shed on the ambiguities, insufficiencies and silences of UNCLOS. Among its undeniable merits is the fact that it has reviewed the physical and geographical characteristics of the straits as a previous question to the determination of the applicable legal regime, a task which has several results, among which is the particular situation which entails few areas in consonance with conventional regulations, together with a possible accumulation of regimes or an absence of legal concepts which are undoubtedly applicable. Furthermore, it contributes the provisions of the de la IMO applicable to the security and control of navigation through the international straits, important information which is a relevant component of the work. There is also a rich legal bibliography although the author does not simply recreate it but uses it selectively for operative not merely speculative or theoretical-objective purposes.

As in her previous books, this task has been carried out by Professor López Martín with the legal precision and meticulousness of her works, while she also achieves a difficult balance between the macrocosm contained in the UNCLOS and the microcosm of each international strait, regardless of whether this is primary or secondary. Thus, the long list of straits is broken down into different types with different characteristics. The final result is an uncommon work within the bibliographical panorama, that is to say, less doctrinal and generalist and with more attention given to the reality of the straits, while remaining loyal to the spirit and finality of the work, which is to show the real state of the question in 2010 and not become bogged down in reminiscences of the 1982 provisions as the number of years elapsed between these dates render this essential. Some years ago, doctrinal interest in international straits was obvious in Spain although this was due to a large extent by the situation of Gibraltar although currently this interest seems to have decreased substantially. However, the objective data and the facts continue as they were apart from the new Spanish geostrategic options. Moreover, global analyses were always carried out within the framework of the bipolar scheme which dominated international relations at the time, and it seems that the perceptible transit from a bipolar situation to the current system with a single super power has reduced interest in the legal regulation of straits when the new situation seems to demand the opposite as the only defence of the weak against the hegemonic nation lies in International Law, and not in the comfortable protection of the friendly super power. A calm and balanced examination from, by and for International Law, while taking into account contradictory State interests, is the concealed unexpressed objective of this work, an objective which has been achieved with no reservations.

University Complutense of Madrid
Madrid, Spain

Luis Ignacio Sánchez Rodríguez

Introduction

For some years, "the acceleration of History" has been felt very strongly in Public International Law, in general, and in International Law of the Sea, in particular, by making changes which have meant a substantial transformation of the existing legislation. The focus of this is undoubtedly the straits used for international navigation which, with the introduction of the *United Nations Convention on the Law of the Sea* of December 10, 1982 – in particular, Part III –, have undergone a complete change as regards the navigation legislation in force, codified in article 16.4 of the *Geneva Convention on Territorial Sea and Contiguous Zone* of April 29, 1958.

In this area, we should point out that the progressive development of International Law in opposition to established rules is normally the result of new interests and needs. The legislation on navigation through the straits used for international navigation is sound proof of this. Strategic and military interests, and air deployment together with the needs arising from the development of commerce and communications are factors which have encouraged, or rather, "imposed", a change in the rules of the game.

There is no doubt that the straits have always played a fundamental role in international life. In fact, without exaggerating, we can even state that one of the most important legal questions concerning maritime navigation has always involved passage through international straits. Thus, it is not surprising that, since ancient times, straits have been the cause of many controversies and, even, wars. Some even believe[1] that the objective of the Trojan War was the interest the Trojans had in controlling the Bosphorus and the Dardanelles (Hellespont). Moreover, no one can deny that Messina, Sicily and Gibraltar played a decisive role in the wars between Rome and Carthage.

However, one thing is certain and that is that the passage through the straits has become more important with the elapse of time as a result of the appearance of political, military and economic interests, which were not new, but were increasingly

[1] Cf. Lapidoth (1972, p.1).

growing stronger. On the one hand, the increase in sea transport has given rise to a greater risk of accidents which affect both the safety of persons and the coastal State as a consequence of the possibility of pollution. On the other hand, the straits have become more significant as regards the supply of petroleum and this has become a more pressing problem; while the straits are crucial for the deployment of air and naval forces in the waters the straits connect, which brings into play the strategic interests of the great powers as regards these sea lanes. Therefore, there is no doubt that the straits are essential sea lanes for all nations; therefore, how their transit must be regulated is strategically vital.

This situation has led to the straits being traditionally considered to be critical points of confrontation between the unquestionable sovereign rights of the coastal States of the straits and the "need" for a more or less justified freedom of movement for the States which use these straits for navigation. This situation affects as to international trade as to the deployment of military power.

The result of this power struggle means that the legislation regarding navigation through straits has been traditionally discussed at "territorialisation" and "internationalisation" level, between the interests of the coastal State and the navigation interests of the naval powers. This classical dual factor in International Law seems to have finalised in the 'theatre' of the III United Nations Conference on the Law of the Sea.

Economic and above all strategic factors have decisively boosted "the revision of the traditional order of things",[2] in order to eradicate the right of innocent passage through international straits. A revision which had been longed for over a long period of time by the Great Sea Powers, headed by the United States and the Soviet Union. The passage of submerged submarines and the free flight of military aircraft through international straits – some of these in particular – were the objectives to be achieved in order to ensure the mobility of the strategic military deployment of the hegemony of the superpowers and their military allies since the right of innocent passage, in force until then, was a hindrance to this factor as it prohibited both submarine and aircraft movement. The coastal States held two important cards: the extension of territorial waters to 12 miles, and the recognition of the 200 miles exclusive economic zone, aspects which were of vital importance for the majority of the international community, and were used as a strategy to put pressure on other nations.

The gauntlet was thrown: "with no straits, no 12 miles and no exclusive economic zone". This attitude meant that it was quite clear that free transit through the straits was necessary for strategic and security reasons in order to facilitate the maximum mobility of the war fleets. The Straits States, with Spain leading them, held fast but did not know how, did not want or could not withstand the attack.

As was expected, the result was the triumph of Goliath over David, that is to say the Part III of the United Nations Convention on the Law of the Sea of 1982: "Straits used for International Navigation", which seemed to put an end to the tortuous road which the legislation on navigation through international straits had

[2]Vid. Jiménez Piernas (1982, p.806).

taken for centuries. A compendium of 12 articles was presented as a sign of the balance between the interests involved in security and the environment of the coastal States and the interests of freedom of navigation of other States. However, we must not be deceived these are in consonance with and favour the strategic, global interests of the Great Naval Powers, to the same extent that they obviate the sacred sovereign rights of the coastal States of the straits and lead to serious damage to their security.

Much has been written about international straits, which is logical considering the importance of the matter although we should point out that all the works on this matter involve two perfectly defined lines of thought. One involves the examination of the legislation on straits in a very general sense, which means that there are many studies which merely explain the legislation in force, and these sometimes refer briefly to the concept of straits used for international navigation through a analyses from a historical perspective, and these are hardly ever illustrated with any examples of practice in any of the straits, except for those which are unanimously considered to be the main straits: Dover, Gibraltar, Malacca, Singapore and Hormuz. The other line of thought is focused on an exhaustive analysis of one or a few specific straits, and includes a description, their legal regime or the problems involved, with no reference to the general regime or the other straits.

Bearing in mind the importance of this issue as regards international straits and the others for some specific States and for all the States, a comprehensive detailed study was required in order to explain and analyse the international regime in force and currently specified in the United Nations Convention on the Law of the Sea de 1982, and to provide a precise, detailed analysis of what should be understood by "straits used for international navigation" so that the subject of application of Part III of the 1982 Convention is perfectly delimited as a categorization of the types of straits identified in the 1982 Convention, including each of the international straits in the world in their corresponding categories. This work is crucial and necessary both for the States and for navigators in order to know which straits have freedom of navigation and flight, which involve the right of innocent passage, which have the right of passage in transit, which are archipelagic sea routes, and which straits do not have any right of navigation as they are subject to the exclusive sovereignty of the coastal State due to the fact that they are interior waters. Besides analysing each and every one of the articles of Part III of the 1982 Convention in detail, the study will complement this analysis with the practice of the States in this regard, including the internal legislation on straits of all and not just some of the coastal States, as well as the question, which is vital for the States and vessels which use these straits, on whether there are traffic separation schemes or other security measures.

Our intention is to make a study of this type in this work.[3] The specific objective is the analysis of the legal regulation of navigation through international straits

[3]However, we should point out that the subject of this work is the legal regulation of straits in pace time, which is the subject of the 1982 *United Nations Convention on the Law of the Sea*. In times of armed conflict, navigation is not governed by the Law of the Sea, but by the Laws of War, which is not examined in this work.

provided by the *United Nations Convention on the Law of the Sea* of December 10, 1982, specifically the Part III, together with the subsequent development of this Convention, especially as regards the practice of the coastal States and the extent of the adaptation of their internal legislation to the new legal regime in order to be able to specify the current situation of the regulation of navigation through international straits.

To achieve this, we start from a brief historical analysis of the legal regime of international straits up to the adoption of the 1982 Convention in order to better understand the problem of navigation through the straits and to find out the points which have changed. This analysis is made in the first two chapters. The first chapter focuses on the most relevant aspects of the regulation of straits before the III Conference, from Grotius, through the first attempts to codify this matter, the decision of the International Court of Justice regarding the *Corfu Channel Case* in 1949, to the work of the International Law Commission which is recorded in the *Geneva Convention on the Territorial Sea and Contiguous Zone* of 1958. The second chapter deals exclusively with the analysis of the development of the *III Conference of the United Nations on the Law of the Sea* as regards the straits used for international navigation from the proposal of Ambassador Pardo in 1967, through the work done by the Commission of the Sea Beds, analysing each of the 11 sessions, the proposals submitted by the States and the texts of the negotiations until the adoption of the *United Nations Convention on the Law of the Sea*, on December 10, 1982.

The exact meaning of the expression "straits used for international navigation", which is not defined in any article of the Convention of 1982, constitutes the subject of Chap. 3. A complex, arduous task due to the many factors which come into play, but one which must be carried out as only those straits which are included in this statement are included in the scope of application of Part III. Related to this first task regarding conceptualisation, in Chap. 4 we untangle the complex catalogue of straits provided in the articles of the Convention, with an explanation of each one of these and the more important and complicated task involving placing each one of over 200 straits in the world in one of the eight categories set out by the Convention. This work has not been done before, but it is fundamental for States and navigators in order to know the legal regime of navigation which governs in each of the straits, and whether they fall within the scope of application of Part III or not as other parts of the Convention of 1982 apply to them, such as Part II, IV, V or VII.

Finally, in Chaps. 5 and 6, a detailed analysis is provided of the two navigation regimes applicable to the "straits used for international navigation", as stipulated in Part III: the classical right of innocent passage, applicable in the secondary straits, that is to say, the straits in articles 38.1 and 45.1(b), which will be analysed in Chap. 5; and the new right of passage in transit, which governs in the main straits in article 37, which are the most numerous and which include the main international navigation routes, which will be the subject of study of Chap. 6. In both cases we will define the modality of passage, specifying who the beneficiaries are, and an analysis will be made of the reciprocal rights and obligations of the coastal States of the straits and the States using them in relation to each modality of passage.

As regards this last aspect, we will endeavour to provide details of the internal legislation of the coastal States of straits concerning the straits, as well as explaining the traffic separation schemes and other security measures adopted in international straits, an aspect which is extremely practical and useful. The last chapter will end with a brief analysis of the possible customary nature of the right of passage in transit.

With the finality to obtain a whole vision of the all straits existing in the world – more or less – and to know their localization and their regime of navigation, a list of international straits, located by geographic areas with the reference of the relevant article in Part III (or Part IV) which establishes the category it belongs to, has been incorporated to the ending of this work.

We wish to draw attention to a final point regarding precision. Throughout this work, we will use the expression "international straits" as the equivalent of "straits used for international navigation" and both expressions are employed. We are aware that this expression is not totally correct as a technical term as it is not used by the 1982 Convention, but we understand that, generally speaking, it expresses the *status quo* of straits under current International Law very well.

Chapter 1
The First Steps in the Regulation of Passage Through the Straits: From Grotius to Pardo

Unlike other parts of the *United Nations Convention on the Law of the Sea* (UNCLOS) of December 10, 1982, such as Part II concerning territorial waters and the adjacent area and Part VII on the high seas, the articles of Part III are not based on any of the Conventions on the Law of the Sea adopted at the First Convention of the United Nations on the Law of the Sea in 1958. However, this does not mean that this part does not have antecedents. Undoubtedly, Part III is the latest attempt among several which have been made by *International Law Practitioners* and Inter-Governmental Conferences in order to establish the legal regulations applicable to international straits as articles.

In this regard, it is evident that the UNCLOS not only introduces substantial progressive development of the subject, but also codifies widely established customary regulations in international legislation. Undoubtedly, both aspects are perfectly combined in the legislation set out in Part III concerning the "Straits used for international navigation", in a symbiotic relationship.

To know how these aspects evolved in practice generally accepted as law up to the Third Conference of the United Nations on the Law of the Sea, as well as to evaluate the most relevant points of the development of this Conference, constitute a *conditio sine qua non* for a proper understanding of the new legislation stipulated in the UNCLOS for navigation in international straits. This legislation is marked by its autonomy as regards the general regime on territorial waters and by the differentiation between several types of straits, and both of these factors entail a radical change in the traditional regulation of this type of maritime area, although it does have certain aspects of continuity as concerns the application of the right of innocent passage which cannot be suspended and it is codified in the *Geneva Convention on the Territorial Sea and Contiguous Zone* of 1958 to certain international straits.

The summary analysis of the most relevant aspects from a historical perspective of the status of straits in International Law up to the 1982 Convention is necessary in order to duly address the detailed study of the legislation of Part III of the UNCLOS on the passage through straits used for international navigation.

A.G. López Martín, *International Straits*,
DOI 10.1007/978-3-642-12906-3_1, © Springer-Verlag Berlin Heidelberg 2010

1.1 The Legal Position of Straits from the Classical Dimension

The issue of navigation through straits "has been of concern since 222 BC, during the Peloponnesian War".[4] Since then the separation of 'freedom' or 'limitation' of passage through these narrow channels has continually involved political, military and commercial interests throughout the centuries. However, it is not until the twentieth century that the idea of specific regulations of International Law arises in order to regulate the regime governing navigation through straits. This delay was probably due to the fact that, until this time, the sovereignty of a State over its territorial waters did not go beyond 3 miles, which almost always left free passage on the high seas through these channels; consequently, the sovereignty of the adjacent shores was not threatened.

However, the lack of specific norms concerning navigation in straits must not be understood as disinterest in the matter. On the contrary, since the beginning of the seventeenth century, Grotius expressly recognised in *Mare Liberum* the right of the owner of the land to occupy gulfs, straits and those parts of the sea which can be seen from the coast,[5] international law doctrine has endeavoured to achieve a consensus regarding a legal regime concerning this, but with little success.[6]

1.1.1 The Most Relevant Traditional Doctrine in This Regard

As we have stated, Grotius was the first to put forward thc extension of state sovereignty over territorial waters, including the straits.[7] However, this right does not empower the occupant to prevent unarmed or innocent navigation through these. Both principles were again formulated more precisely in his treatise *De Jure Belli ac Pacis*, with the addition of a third grand principle: the right of the occupant of one of these portions of the sea – gulfs, straits and coastal sea – to impose equitable charges on sailors if it took "upon itself the responsibility of defending navigation and providing assistance through night time fires and short signals".[8] This last principle was in force up to the middle of the nineteenth century, when it was eliminated as a consequence of Denmark signing some international treaties with European States and with the United States, whereby the ships of the other nations were released from paying levies to pass through the *Belts Straits* and

[4]Cf. Wainwright (1986, p. 366).

[5]Vid. Grotius (Trad. V. Blanco García and L. García Arias) (1956, p. 16).

[6]A detailed study of the historical evolution of the regime of straits, from the point of view of doctrine and state practice, can be found in the excellent work of Brüel (1947, pp. 48–205).

[7]Przetacznik (1977, p. 223–225) even goes back to the era of the Phoenicians in order to find claims over sea areas adjacent to the coast. For obvious reasons we will not go into a study of these remote antecedents; his analysis of these can be found in his work.

[8]Vid. Grotius (trad J. Torrubiano) (1925, p. 324).

the *Sund*.[9] This is a rough outline of principles which subsequent doctrine will make more precise.

However, for Grotius the question of the straits constituted only a small part of a more important matter, freedom of the seas. The first author to take a direct interest in the straits was Pufendorf who expressed the need for the coastal state to secure its ports, as well the need to control the part of the straits nearest to its coast. Like Grotius, he recognised the right of the coastal State to charge a toll to pass through the straits.[10]

However, a century latter, Vattel was the first to distinguish between 'straits which serve as a means of communication between two seas where navigation is common to all or many nations' and 'straits which do not have this function'. As regards the former, the nation which 'possesses' the strait cannot prohibit the passage of other nations through the straits, on condition that this passage is innocent and does not entail any risk for the security of the coastal State. However, the coastal State of the straits is authorised to take certain precautions and to request certain formalities insofar as these affect its security. Like Grotius and Pufendorf, he conditions passage through the straits to the payment of a moderate levy. Vattel bases the right of navigation of foreign ships through territorial waters, including straits, on the concept of 'right of way' in Roman law.[11] In fact, the majority of authors of the eighteenth century defend the right of the coastal State to exclude the navigation of foreign ships from its territorial waters.

These interests of the coastal State put forward by Vattel became secondary in the nineteenth century as opposed to more liberal notions of right of way through straits, which is conceived as a natural complement to the freedom of the seas. This change of orientation was related to the emancipation of the Spanish colonies in America, which involved the consolidation of the 'open doors policy' for trade between Europe and America, with an evident repercussion on the conception of the regime regulating the passage through straits.

The nineteenth century saw the beginning of the recognition that straits must be subject to a special regime, and the point was raised that, although waters are territorial for some purposes, for purposes of transit, they are equivalent to the high seas.[12] In addition, some authors such as Holland make a distinction regarding the conditions of the navigation of straits in peacetime and in wartime; in the first case, the passage of all vessels including warships must be considered to be innocent; while in the second case, the coastal State can deal with enemy ships as it may wish.[13]

[9]On March 14, 1857, Denmark signed treaties with Prussia, Austria, Belgium, France, England, Hanover, Mecklenburg-Schwerin, Oldenburg, the Netherlands, Russia, Sweden and the Hanseatic Cities of Lübeck, Bremen and Hamburg; on April 14, 1857, with the United States, and February 25, 1860 with Spain. Vid. the quotation in Meseguer Sánchez (1999, p. 156).

[10]Cf. Pufendorf (trad. J. Barbeirac) (1712, pp. 331–333).

[11]Cf. Vattel (trad. M.M. Pascual) (1834, p. 273).

[12]This is an approach of Godey, undoubtedly, the first author to recognise such a *sui generis* regime (cf. Brüel 1947, p. 59).

[13]Cf. Holland (1898), op. cit. in Brüel (1947, p. 60).

This liberal tendency in the interpretation of the right of way took on more importance in the twentieth century. The most extensive study of this period concerning the problems inherent to international straits was the one put forward by Schücking. This author stresses that not all the straits which connect two high seas areas are necessarily of interest for International Law, but only those which are of value for international navigation. In addition, he sustained that those straits which have high seas corridors within their half, the general rules concerning the high seas should be modified in order to restrict any war manoeuvres in this area.[14]

To sum up, the abundant literature from the classical era on navigation through straits generally tends to affirm the right of way of all vessels – both merchant ships and warships – in peacetime, while in wartime, there is almost total unanimity regarding the recognition that only merchant ships of the States which are not at war with the coastal State have right of way, unless it is a neutral coastal State as such a State cannot close its straits.

1.1.2 The First Attempts at Private Codification About the Legal Regime of Straits

In addition to the attention paid by classical doctrine, several private scientific associations have subjected the question of the legal status of international straits to examination in an attempt to systematise a group of specific rules.

In this sense, the work done by the *Institut de Droit International (IDI)* from 1894 to 1912 made decisive progress towards the formulation of autonomous legal regime for straits. In fact, the proclamation of the principle of innocent passage in straits by the IDI in 1894 was the key to the creation of a special category of straits in which the territorial rights and the rights of way converge. This formulation was accompanied by the definition of straits as those which do not "exceed double the width of the territorial sea".[15] The waters of these straits would be territorial waters, but, due to the fact that they serve as a passage from one area of free sea to another, these straits can never be closed and transit through these must be free.

In the debate on the laying of mines in 1906, it was proposed that neutral States did not have the right to lay mines in the straits which link high seas with high seas.[16] In the Paris Session in 1910, the IDI considered the passage of warships through the straits in relation to the right of neutrality. Some members were in favour of the right of way, others put forward the limiting this to one strait if there

---

[14]Cf. Schücking (1906), cit. in Brüel (1947, p. 65).

[15]Vid. *Annuaire del'IDI*, 1894, vol. III, p. 393. For a detailed analysis of the debates and proposals of the IDI from 1894 to 1912 see Brüel (1947, pp. 70–80).

[16]Cf. *Annuaire de l'IDI*, 1906, vol. V, p. 275.

were alternative channels, and others suggested limiting this to the 'essential' straits. The Drafter attempted to combine these ideas in one commitment formula, according to which the coastal state would not violate the right of neutrality in the event that it might decide to prevent the passage of belligerent warships on condition that this is not applied to maritime routes necessary for navigation; that is to say, it had to permit passage in those straits which constituted the only means of communication from one free sea to another. This same distinction again arose in the debate on naval warfare in the Oslo session of 1912, when it was decided to prohibit the laying of automatic mines in straits; although it was suggested that this limitation would affect those straits which were the only or essential channel of communication between areas of the high seas.[17] However, the articles project was finally brought to a halt due to the beginning of the First World War.

During this same period of time, the subject of straits was also addressed by the *International Law Association (ILA)* from 1894 to 1910. Thus, the ILA meeting held in Brussels in 1895 adopted the essential norm for communications which was that straits could never be closed.[18] In 1906, the ILA agreed that belligerents could not block straits to the extent that this would interfere with navigation. In this regard, it "Rules on Territorial Waters" adopted in 1906 contained the following provisions on straits:[19]

> ARTICLE 10.- Les dispositions des articles précédents s'appliquent aux détroits dont l'écart n'excède pas douze milles, sauf les modifications et distinctions suivantes:
>
> 1. Les détroits dont les côtes appartiennent à des États différents font partie de la mer territoriale des États riverains, qui y exerceront leur souveraineté jusqu'à la ligne médiane.
> 2. Les détroits dont les côtes appartiennent au même Ètat et qui sont indispensables aux communications maritimes entre deux ou plusieurs États autres que l'État riveraise ont toujours partie de la mar territoriale du riverain, quel que soit le rapprochement des côtes. Ils ne peuvent jamais être barrés.
> 3. Dans les détroits dont les côtes appartiennent au même État, la mer est territoriale bien que l'écartement des còtes dépasse douze milles, si à chaque entrée du dètroit cette distance n'est pas dépassée.
> 4. Les détroits qui servent de passage d'une mer libre à une autre mer libre ne peuvent jamais etre barrés.
>
> ARTICLE 11.- Le régime des détroits actuellement soumis à des conventions ou usages speciaux demeure réservé.

The position of the IDI and the ILA concerning the status of straits can be summed up in the following terms. With regard to the special rules which must be formulated for straits, both institutions highlight the importance of trade in straits. The question of the extension of territorial waters in straits which are of international importance arises in relation to the fact that the international straits must not exceed double the width of the territorial sea; and in those cases in which both

[17] Cf. *Annuaire del'IDI*, 1912, vol. VI, pp. 697–702 and p. 78, respectively for each session.

[18] Cf. *Report of the 17th Conference*, p. 102.

[19] Vid. *Report of the 23rd Conference*, p. 251, cit. in Brüel (1947, p. 80).

coasts of the straits belong to the same State, its right to consider the totality of the straits as territorial waters is recognised.

As regards conflict of interests between belligerents and neutral coastal States in straits, it is evident that the tendency in the debates of both Institutes is towards the limitation of the right to war to the benefit of the right to peace, and to place freedom and security on the sea above the interests of the belligerents; at the same time, it was maintained that the neutral coastal State could not obstruct the passage of warships belonging to belligerent powers.[20]

However, none of the recommendations of the IDI and of the ILA were incorporated into an international agreement.

These same theories were considerably extended throughout the debates of the *Inter-parliamentary Union* held from 1910 to 1915. They were dealt with in more detail and developed in a way which was more in accord with the practical evolution of the States, as was to be expected from an organism composed of politicians and practitioners. Its proposals were recorded in a Project of Agreement which should have been submitted to discussion for approval at the Conference planned for August 1915 in Stockholm.[21] The breakout of war also brought this project to a halt.

Finally, as concerns doctrine and the scientific and political associations we have referred to,[22] we can conclude from the options given that the important role of international straits required that they become the subject of a *sui iuris* legal status different from what was stipulated for other maritime areas. Nevertheless, this same retrospective view of these opinions leads to stating that these do not constitute a sufficient basis to create positive law in this regard.

1.1.3 The Incipient 'Official' Steps Towards the Regulation of Passage Through the Straits

As we have seen, the International Law doctrine was unanimous as regards the need to structure a body of autonomous norms which would regulate navigation through international straits. However, state practice fell very short of satisfying the need; the fact was that each strait had its own particular regulation, and there was no common regulation. At the beginning of the twentieth century, there occurred the first 'timid' attempts to draft a group of International Law norms in this regard,

---

[20]Cf. Brüel (1947, p. 80).

[21]For a study of all the debates of the de Inter-parliamentary Union as well as a detailed analysis of its project of agreement, see Brüel (1947, pp. 81–87).

[22]Besides the institutions referred to the straits issue was also debated by the American Institute of International Law (1926) – although this had scarce repercussion –, and by the Japanese Association of International Law (1926).

although it should be taken into account that the Peace Conference held in The Hague in 1899 did not address any question concerning maritime areas.

1.1.3.1 The Peace Conference of The Hague in 1907

The question of the legal status of straits was first discussed at the inter-governmental Conference held in The Hague in 1907. In this Second Peace Conference, the question of the straits arose in connection with the *VIII Convention concerning mine laying*, and the *XIII Convention on the obligations and rights of neutral States.*[23]

During the debate in the first of these Conventions, the Netherlands submitted a proposal to exclude straits from the right to lay mines. When this was referred to the Examining Committee, no one objected to the obligation to leave a passage in the Straits which would join two free seas, but the modification that the passage should be in accordance with the conditions stated by the competent authorities was also included. In fact, the provision on straits was finally omitted.

Sweden brought up the question of straits at the XIII Convention. There was agreement concerning the fact that, while passage could be prevented through territorial waters in order to preserve the neutrality of the coastal State, this was not applicable to straits which joined two areas of the high seas.

The only clear conclusion that could be drawn from this Conference was the need for separate codification for straits.

1.1.3.2 The 1930 Hague Conference on Codification

The Conference on Codification of International Law held in The Hague in 1930 and convened by the League of Nations[24] mean a new attempt in the process of codification after the failure of the attempts at private codification which were halted due to the breakout of war.

Forty-seven States sent delegates to The Hague, seven of which, including the United States, were not members of the League of Nations. Deliberations lasted three weeks from March 13 to April 11, 1930.

One of the essential points examined by this Conference in The Hague was the extension of territorial waters. On this point, the Conference was meticulously prepared on the basis of *ad hoc* studies, and questionnaires sent to the participating States. The responses related to the width of the territorial sea showed the disparity

---

[23]Cf. O'Connell (1984, pp. 302–303).

[24]In 1927, the Assembly of the League of Nations decided that codification would focus on three subjects: (1) Nationality; (2) International Responsibility; and (3) Territorial Waters. The norms concerning straits were included within the framework of this last subject. For a detailed analysis of the debates and proposals concerning the 1930 Hague Codification Conference, see the interesting work of Brüel (1947, pp. 175–195).

of opinions on the matter, which made it impossible to reach an agreement on the width of territorial waters and on the exercise of special jurisdiction in the zone adjacent to this area.

However, within the framework of the discussion on these matters, the Preparatory Committee of the Conference managed to direct the debates toward addressing the question of the straits, bearing in mind that the situation of straits had not been included as a specific subject in the Discussion Bases of the Preparatory Committee of the Conference, despite the fact that the principle of freedom of passage and navigation through the Turkish straits (*Bosphorus*, *Dardanelles* and *the Sea of Marmara*) had already been recognised in the Treaty of Lausanne in 1923, and in the Montreaux Convention of 1936 concerning the regime on straits. Thus, straits were addressed within the context of the delimitation of territorial waters and concerned the delimitation of the straits. Navigation through the straits in which the territorial waters of two coastal States overlapped was not considered as a subject distinct from the question of innocent passage through territorial waters in general. The most relevant aspects in this codification process were as follows.

The Security Council of the League of Nations appointed a Preparatory Committee to draft a report which would serve as a basis for the subsequent work of the Conference, together with the responses from the governments. This was prepared by Schücking. The report contained that the regime on straits subject to special agreements must be excluded from codification; it considered that the waters of the straits were territorial waters if the entry to the straits did not exceed 10 miles; and established the half way line as the method for delimiting the straits which did not exceed 10 miles or whose coasts belonged to several Sates.

However, this report complicated the analysis of the straits as it distinguished between those waters which were territorial from those which were interior, based only on the fact that there was several or only one State involved. In the case of straits whose coasts belong to more than one state, the principles related to territorial waters apply, while, in the case of straits whose coasts belong to one State, if the width does not exceed the double of the territorial waters, the principle of bays is applied. In the first case, there would be a right to innocent passage; in the second case, there would be no right of passage. Finally, the Report considered that the rule that straits can never be closed, in accordance with the idea that a coastal State has the right to completely close its territorial waters in wartime.

The questionnaire submitted to the Governments by the Preparatory Committee was even more enigmatic. The Governments were requested to comment on the conditions which determine which territorial waters in straits connect two areas of open seas, or the open sea with an interior sea when the coasts belong to one or several States. The majority of the responses were not inclined to the characteristics of a right of passage through the straits, but referred to the dimensions of the territorial waters in the straits. Among the few responses which referred to the right of passage, the response of Germany should be stressed as it stated that territorial waters did not affect the right of innocent passage through straits for all kinds of vessels, and underlined the need to extend this right to aircraft flying over straits, but only those with a number of coastal States.

When examining the Final Report of the Preparatory Committee and the responses of the Governments, three Committees were created in order to consider each of the three Discussion Bases. The width of the territorial sea corresponded to the Second Committee, within which the specific subject of 'straits' was assigned to the Second Sub-Commission of the Second Committee.

In its report to the Principal Committee of the Conference, the Second Sub-Commission abandoned the distinction between straits which belong to one or several States, and replaced this with the distinction between straits which join two parts of the high seas and straits which only provide access to interior waters. A reference was added to the Project of Convention – made up of 13 articles – prohibiting the coastal State with straits which joined two parts of the high seas and are used for international navigation from preventing the passage of warships under no circumstances. However, when The Hague Codification Conference put forward the question of passage through straits, this was done vaguely and only in the context of the right of passage of warships through territorial waters. No express provisions were made as regards straits.

No agreement was reached. Only the statement of the Second Committee remains. This stated that, in those waters of straits which constitute territorial waters, "it is essential to ensure, in time of peace in all circumstances, the passage of merchant vessels and warships through straits between two parts of the high seas forming ordinary routes of international navigation".[25]

As regards the Spanish position at this Conference, the Spanish delegation presented an amendment to the basis of discussion number 19 on the right of innocent passage, specifically related to warships. There are three outstanding points: navigation must be on the surface, it was prohibited to discriminate between the flag of the coastal State and the foreign flags or between the foreign flags, and that the right of innocent passage can only be abolished for reasons of public security and in cases of serious events which might affect the security of the coastal State or essential interests of this State.

In conclusion, while the previous reports, the government responses, and the debates of the Second Committee have a certain value and cast light on the questions concerning straits, the same cannot be said of the final result of The Hague Codification Conference in 1930. In fact, as was stated in the United Nations, "the treatment of the question of straits in all of these exercises between 1894 and 1930 was unsatisfactory".[26]

[25]Vid. *Minutes of the Second Committee*, 1930, p. 148; document cited in Anand (1980, pp. 111–112).

[26]Office for Ocean Affairs and the Law of the Sea (1992, p. 2).

1.2 The Desired Concretion of the Legal Regime on Straits

Ambiguous and imprecise are the adjectives which best describe the legal regime on navigation through international straits until well into the twentieth century, as we have been able to verify above, with the exception of some specific straits, such as those in Turkey, Denmark and Magellan Straits, in which the passage of vessels was totally or partially regulated by particular agreements.[27] This is a situation which persisted until the International Court of Justice (ICJ) became involved in the matter.

In fact, the decision dictated by the Court in 1949 in the *Corfu Channel Case,* one of the first cases submitted to this international judicial authority, entails the confirmation of a specific singularity for these channels of communication, which would be conventionally enshrined in the *Geneva Convention on Territorial Sea and Contiguous Zone* of 1958, the result of the work of codification and progressive development carried out by the United Nations International Law Commission.

1.2.1 The Cornerstone for Settling the Regime on Straits: The 'Corfu Channel Case'

On May 15, 1946, two British cruisers were sailing through the North Channel of the Strait of *Corfu,* which separated the Greek island of Corfu from the Albanian mainland, and came under Albanian artillery fire. The warships were undamaged and did not return the fire, but the United Kingdom protested. The United Kingdom claimed that the waters were open to international traffic while Albania stated that the waters of the straits belonged to it and that passage through these straits required permission from Albania. After some diplomatic correspondence, the United Kingdom affirmed its right to send warships through the straits and warned that if these were fired upon, they would return the fire.

Some months after the first incident, on October 22, 1946, a British squadron made up of the cruisers *HMS Mauritius* and *HMS Leander*, and the destroyers *HMS Saumarez* and *HMS Volage*, crossed the north of the Straits of *Corfu* on a routine mission. Near the Bay of Saranda, first the *Saumarez*, and then the *Volage* were struck by mines which had been laid in the straits, exploded, and were seriously damaged; 44 naval personnel were killed and another 42 were seriously injured. On November 13, 1946, the North Channel of *Corfu* was swept by several minesweepers of the Royal Navy which found another 22 mines.

---

[27] These straits regulated by special agreements will be the subject of more attention in Chap. 4 of this book when reference is made to those straits which remain beyond the regulation of Part III of the UNCLOS; specifically, as regards what is stipulated in article 35(c), which excludes the straits regulated by *long standing* agreements still in force from this legal regime, such as the Straits in Turkey, Denmark and the Straits of Magellan.

In a Resolution of April 9, 1947, the Security Council of the United Nations recommended that the two States involved (the United Kingdom and Albania) submit their dispute to the ICJ. The United Kingdom filed a written request at the Court Registry, based on article 40.1 of the Statute. Albania answered this request before the ICJ, arguing that it was not in consonance with what was stipulated by Security Council which required an agreement by both parties in order to submit the case to the Court. On March 25, 1948, the ICJ rejected the objection of Albania to its jurisdiction, by 15 votes to 1. On that same day, both States submitted the case to the Court through a Special Agreement.

The United Kingdom[28] accused Albania of infringing the right of innocent passage through straits which connect two parts of open seas, as well as the specific rules on mine laying established by The Hague Convention of 1907.

Albania argued[29] that, from the time that a state of war existed between Albania and Greece, the passage of foreign vessels could not be considered to be a simple question of freedom of passage, but, in these circumstances, it should be considered to be a matter of national security. In addition, Albania held that there were two different classes of straits, those which constituted a necessary route between two parts of the high seas and are of interest for the international community due to international trade, and those whose transit is not essential for the international community. For Albania, there was no doubt that the North Channel of *Corfu* constituted strait in a the geographical sense, but it denied that this channel belonged to the category of international maritime channels through which there was a right of innocent passage. It considered that this channel was only of secondary importance, which was used exclusively for local traffic, and was not a necessary route between two parts of the high seas.

In its decision of April 9, 1949, the ICJ stipulated the following:

> It may be asked whether the test is to be found in the volume of traffic passing through the Strait or in its greater or lesser importance for international navigation. But, **in the opinion of the Court, the decisive criterion is rather its geographical situation as connecting two parts of the high seas and the fact of its being used for international navigation**. Nor can it be decisive that this Strait is not a necessary route between two parts of high seas, but only an alternative passage between the Aegean and the Adriatic Seas. It has nevertheless been a useful route for international maritime traffic...
>
> Having regarded to these various considerations, the Court has arrived at the conclusion that the North Corfu Channel should be considered as belonging to the class of international highways through which the passage cannot be prohibited by a coastal State in time of peace.[30]

Thus, in the opinion of the Court, the decisive criteria for the application of the right of passage through straits were the geographical situation that the straits

---

[28]Cf. *The Corfu Channel Case. Pleadings, Oral Arguments and Documents*. ICJ *Reports*, 1949, vol. I, p. 21.

[29]Cf. *The Corfu Channel Case. Pleadings, Oral Arguments and Documents*. ICJ *Reports*, 1949, vol. III, p. 383.

[30]Vid. ICJ *Reports*, 1949, pp. 28–29 (letters in bold are ours).

connected two areas of the high seas, and that they were used for international navigation.

Albania also argued that, in the case of classification as international straits, the passage of warships was conditioned to previous notification; therefore, the United Kingdom had infringed Albanian sovereignty when it sent its warships with no previous authorisation of Albania. In this regard, the Court held that:

> It is, in the opinion of the Court, generally recognized and in accordance with international custom that States in time of peace have a right to send their warships through straits used for international navigation between two parts of the high seas without the previous authorization of the coastal State, provided that the passage is innocent. Unless otherwise prescribed in an international convention, there is no right for a coastal State to prohibit such passage through straits in time of peace.[31]

Concerning mines, it argued that there was no evidence relating this with the incident, nor was there any evidence that this was known, therefore, the principles of The Hague Convention of 1907 were not applicable, an allegation rejected by the Court.[32]

The decision in the *Corfu Channel Case* was presented[33] as a "judicial victory" and a "naval failure". It was a victory as it recognised the right of passage of warships with no authorisation, contrary to the intentions of the Albanians; a failure as the British demonstration had required substantial resources (an aircraft carrier, cruisers and other warships) and the requirement that its conduct be 'innocent' placed it in a position of relative inferiority as regards the coastal States.

In synthesis, we can conclude that the decision of the ICJ in April 9, 1949 introduced the following specifications as regards the regime of innocent passage through international straits:[34]

- Warships have the right to innocent passage.
- This right cannot be suspended or conditioned by the requirement for previous authorisation by the coastal State of the straits when the straits are used for international navigation between two parts of the high seas.
- The defining components of international straits are the geographical component involving the geographical component of connecting two parts of the high seas and the functional component involving its use for international navigation.
- In order to classify straits as 'used for international navigation', it is sufficient that the straits in question are a useful route for international maritime traffic, they do not have to constitute a main route.

Nevertheless, some members of the Court were not fully in agreement with this decision, and held that there was no right of innocent passage in International Law. The dissident opinion of Judge Krylov illustrates this philosophy when he stated:

---

[31]Vid. *Ibidem*, p. 28.

[32]Cf. *Ibidem*, pp. 11–12 and p. 29.

[33]Cf. O'Connell (1975, p. 107).

[34]Cf. Yturriaga Barberán (1991a, p. 30).

> There is no common regulation of the legal regime of straits. Every strait is regulated individually... by convention. The regime of the Corfu strait has not been judicially regulated. If the regime of the strait is not defined by a multilateral convention, it appertains to the coastal State or States to regulate it.[35]

1.2.2 The Work of the International Law Commission and the United Nations Conferences on the Law of the Sea Held in Geneva

The decision of the ICJ in the *Corfu Channel Case* clarified certain aspects of law which regulated passage through international straits. However, for reasons of legal security, it was necessary to codify this matter, just as the early doctrine and the States had unsuccessfully attempted to do since the nineteenth century. This background and, especially, the 13 articles on territorial waters prepared by the Second Codification Conference of The Hague of 1930, constitute important material for the work of codification carried out by the International Law Commission (ILC), brought about in the hope of rendering the incipient legal regime more specific.

1.2.2.1 The Achievement of Consensus: The Geneva Convention on Territorial Sea and Adjacent Areas of 1958

The key or 'critical' date for the commencement of consensus regarding a number of rules which govern navigation through international straits was 1949. At the same time as the ICJ was dictating its important decision resolving the *Corfu Channel Case*, the ILC began its work in the area of the Law on the Sea, and chose the regime on territorial seas and the high seas as subjects which required codification.[36] However, considering that the problems concerning territorial waters, the adjacent zone, the continental platform and the high seas were closely linked to each other, both physically and legally, in 1953, the General Assembly decided not to begin to deal with these aspects until the ILC had analysed them all and issued a Report.

---

[35] Vid. ICJ *Reports* (1949, p. 74).

[36] For a detailed analysis of the work of the ILC Law of the Sea and the development of the First Conference of the United Nations on the Law of the Sea in 1958, see the publication of Office for Ocean Affairs and the Law of the Sea (1992, pp. 4–20); as well as the important study of Caminos (1987, pp. 44–56). Also of interest are the works of Koh (1982, pp. 35–47); Lapidoth (1972) and Yturriaga Barberán (1991a, pp. 31–37), among others.

In 1953 and in 1955, two projects were sent to the governments so that they might comment on them; only 25 States responded. Finally, in 1956 the Commission completed its definitive projects.

In its Final Report in 1956, corresponding to its eighth session, the ILC recommended that a Conference of Plenipotentiaries be convened in order to examine the legal, technical, biological, economic and political aspects of the Law of the Sea. It also understood that this conference should be carefully prepared by a Project of Articles which appeared as an annex to its Report.[37] In February 21, 1957, by Resolution 1105 (XI) the General Assembly approved requesting the Secretary General that he convene the First Conference of the United Nations on the Law of the Sea, stipulated for the beginning of 1958 in Geneva. On April 29, 1958 the four Geneva Conventions were adopted: territorial sea and adjacent area; continental platform; high seas and high sea fishing.

The question on navigation through international straits was addressed within the general regime of navigation through territorial waters. A circumstance which was undoubtedly surprising if we take into account that all the previous doctrine, which we have referred to, unanimously claimed the creation of an autonomous regime of passage through straits. In the same way, attention should be drawn to the treatment which the ILC granted to the straits where it repeats the classification used in the formula of The Hague Codification Conference of 1930 and other similar formulas based on the mere 'use' of the straits, creating a complete abstraction of the patterns of classification established by the Court in 1949, for which straits are a legal and not merely a geographical institution.[38]

The question of the straits also arose within the ILC in two different contexts. In connection with innocent passage in general and in relation to the delimitation of the territorial waters of States with coasts opposite each other.[39]

In the first case, attention was drawn directly to the straits because it as suggested[40] that, if the straits are subjected to one rule of innocent passage, this circumstance could result in treating all the straits as identical, when there are three categories in fact: those which are subject to international agreements, such as the Bosporus; those which, although they are not regulated by any international agreement, are important for international navigation; and those straits used for international navigation. This leads to noting[41] that the difference between innocent passage in territorial waters used for international navigation is that, in the first case, the warships do not have absolute right of passage, while in the second case

[37]Cf. ILC, "Report to the General Assembly Covering the Work of its Eighth Session", *GAOR*, 1956, Supl. no. 9.

[38]This 'peculiar' circumstance, undoubtedly, had much to do with the fact that the special Drafter of the ILC in this respect was Professor Françoise, who had already been the Drafter of the Second Sub-Committee of The Hague Conference in 1930.

[39]As regards both these questions, see O'Connell (1982, pp. 314–316).

[40]This is the suggestion of Zourek (*ILC Yearbook*, 1955, I, p. 150).

[41]This proposal was put forward by Fitzmaurice (*ibidem*, p. 150).

they do. This proposition connects with the suggestion[42] that the coastal State could prevent warships suspicious of acting against the Charter of the United Nations passing through their straits.

When the ILC considered demanding authorisation or notification of the intention of warships to pass through territorial waters, it then became really necessary to treat straits as autonomous, and legally define them. The decision was taken to identify them by reference to their 'use' in international navigation, rather than due to their being 'essential'.[43] The text which appeared transformed straits into an exception in the case of warships in territorial waters.

Subsequently, there arose the question of straits which do not join two parts of the high seas, but lead to internal waters belonging to coastal States. Discussion took place on the convenience of applying the regime on bays to these straits, and it was proposed that the requirement formulated in the *Corfu Channel Case* concerning the fact that straits connect two parts of the high seas be modified and that the definition of straits be reopened in order to value their usefulness to maritime traffic.[44]

Besides this debate, a proposal was put forward for a new provision concerning straits whose coast belonged to one State. When the project of the ILC was sent to the governments for comments in 1955, Israel realised what this meant for the Gulf of Aqaba and claimed that, regardless of their location, the straits which constitute the only access for a port belonging to a State must not come under the territorial waters regime of another State, which could decide on the suspension of innocent passage.[45]

However, the ILC decided not to include this second class of straits and limited itself to including what was established by the ICJ in the *Corfu Channel Case*. Likewise, no express reference was made to the innocent passage of warships. Finally, article 17, paragraph 4 of the Project on definitive articles on territorial seas drafted by the ILC was drawn up as follows:

> There must be no suspension of the innocent passage of foreign ships through straits normally used for international navigation between two parts of the high seas.[46]

From February 24 to April 27, 1958, the *First Conference of the United Nations on the Law of the Sea*[47] took place. During this Conference, this paragraph of the project of the ILC was adopted, but with two significant changes: the suppression of

[42]A suggestion made by Scelle (*ibidem*, p. 151).

[43]Zourek submitted a motion in favour of the 'essential' nature of straits, which was rejected (cf. *ibidem*, p. 151).

[44]Cf. *Ibidem*, p. 98.

[45]Cf. *ILC Yearbook*, 1956, II, p. 59.

[46]Vid. ILC, "Report to the General Assembly...", *op. cit.*, p. 273.

[47]The position adopted by the Spanish Delegation at this Conference was similar to the one adopted at the Codification Conference of The Hague in 1930 (vid. *supra* note 19), as can be deduced from the proposals presented concerning articles 24 and 25 of the definitive project of the ILC (cf. Doc. A/CONF.13/C.1/L.36).

the word 'normally' and the addition of a new category of straits in which the right of passage cannot be suspended.

The suppression of the term 'normally' was proposed by the United States, which argued that "the Court's opinion did not qualify this phrase by use of the word *normally*".[48] This proposal was withdrawn in favour of a more substantial amendment co-sponsored by the Netherlands, Portugal and the United Kingdom, which included the suppression of the word 'normally', together with the addition of a new kind of straits.[49] This involved the fundamental distinction between straits which join two parts of the high seas and straits which lead to interior waters, and led to most discussion among the delegations at the Geneva Conference. Both promoters and detractors considered that this amendment was specifically aimed at the Gulf of Aqaba and the Straits of *Tiran*. It was approved by the First Committee of the Conference by a narrow margin of 31 votes in favour, 30 against and 10 abstentions. At the plenary meeting, the article was passed with 60 votes in favour and 9 abstentions.[50]

Thus, the only *ad hoc* provision concerning navigation through straits is found in Section III of the Geneva Convention on the Territorial Sea and Contiguous Zone of 1958, within the "Innocent Passage in the Territorial Sea". This is article 16, paragraph 4:

> There must be no suspension of the innocent passage of foreign ships through straits normally used for international navigation between two parts of the high seas, or a part of the high seas and the territorial sea of a foreign State.[51]

Thus, the fact that international straits are a part of the territorial sea is enshrined in law, "precisely that part in which the right of innocent passage as a commitment of the interests of the coastal State and the International Community as regards navigation takes on its authentic, unquestionable significance".[52]

This means that, except for the possibility of suspension, the regime on navigation in straits is characterised by the same right of innocent passage as in the rest of the territorial sea (article 14); that is to say, all the ships which might cross the straits while they navigate through territorial waters and do not enter interior waters, or move towards interior waters or come from interior waters in order to sail towards the high seas; this passage includes stopping and anchoring on condition that these are ordinary navigation incidents or are due to force majeure. The

---

[48]Vid. Doc. A/CONF.13/C.1/L.39, of March 25, 1958.

[49]Cf. Doc. A/CONF.13/C.1/L.71, of March 28, 1958.

[50]This is due to the fact that the complete article was submitted to a vote, a motion to vote paragraph 4 separately put forward by Saudi Arabia was rejected by 34 votes against 32 and 6 abstentions. The representative of Saudi Arabia, Shukairi, declared that "the amended text no longer dealt with general principles of international law, but it had been carefully tailored to promote the claims of one State" (vid. *UNCLOS I*, 1958, vol. II).

[51]We must point out that this text differs from the formula of the *Strait of Corfu*, as it refers to the "suspension" of passage and not to "prohibition".

[52]Vid. Remiro Brotóns (1975, p. 475).

passage is innocent insofar as it does not harm peace, public order or the security of the coastal State, and it will not be considered innocent if the laws and regulations laid down by the coastal State to prevent boats fishing in its territorial waters are not complied with. The coastal State cannot hinder the innocent passage and is obliged to publicise any danger to navigation it may know of (article 15).

As regards the transit of submarines, the project of the ILC only laid down that these must navigate on the surface, but the Conference accepted a French amendment[53] which required the clear hoisting of the flag. Therefore, submarines must navigate on the surface and show their flags while they cross the straits (article 16.6).

This right of passage does not include aircraft. The question of navigation was examined by the ILC, including the special Drafter who was interrogated regarding the possibility of extending the right of innocent passage in the air space. The response of François was definitive, insofar as the *Chicago Convention on Civil Aviation* of December 7, 1944 did not recognise the right of innocent passage, and that the 193 Codification Conference of The Hague had expressly stated that there was no customary right of innocent passage over the air space of the territory.[54] Along the same lines, the OACI pointed out in its comments on the Project of the ILC that any foreign aircraft did not have any right of passage over the terrestrial superficies of a State or its adjacent territorial sea.[55]

In the light of the above, the ILC does not include any provision regarding aircraft flying over the straits. In fact, the only provision in the Geneva Convention on Territorial Sea and Contiguous Zone of 1958 as regards aerial matters is the article 2 which recognises the sovereignty of the coastal State over the air space above its territorial sea.

There is no doubt that the assimilation of the regime of the straits as regards territorial waters perpetrated by the Geneva Convention of 1958 constitutes a step backwards as regards the creation of a special regime for international straits.

1.2.2.2 The Second Conference of the United Nations on the Law of the Sea

The debates of the First Conference of the United Nations on the Law of the Sea showed that there were important questions on the Law of the Sea pending, such as the width of the territorial sea or the creation of a fishing zone, either because these were not included in the Project of the Articles of the ILC, or because the required agreement of the States is not achieved. Due to this, a final resolution was adopted, in which the General Assembly was requested to study the following session, "the

[53]Cf. Doc. A/CONF.13/C.1/L.6, of March 14, 1958.

[54]Cited in Yturriaga Barberán (1991b, pp. 362–363). This opinion was shared by Professor Pepin, who affirms that the movements of aircraft in both areas are subjected to identical conditions given that the air space over the territorial sea is totally assimilated to the air area over the terrestrial superficies (cf. Doc. A/CONF.13/14, of October 4, 1957).

[55]Cf. Doc. A/CONF.13/31, of June 24, 1958.

advisability of convening a second international conference of plenipotentiaries for further consideration of the question left unsettled".[56]

Resolution 1307 (XIII) of the General Assembly approved the convening of the Second Conference of Nations on the Law of the Sea in order to address the breadth of the territorial sea and the fishing zones. This Conference took place from March 17 to April 26, 1960, in Geneva. Up to this time, 22 States had established the width of their territorial waters at 3 miles, 16 had established between 4 and 11 miles, 13 States had established at 12 miles and 2 States over 12 miles.[57] However, despite the fact that all the proposals of the States on the width of the territorial sea exceeded 3 miles, none of these achieved the required consensus.[58]

None of the proposals submitted referred to the international straits although, during the final discussions of the Second Conference, the sea powers made it clear that a regime of unrestricted passage for ships and submarines through the straits, as well as the freedom for aircraft to fly over these was essential, not only regarding the extension of the territorial sea to 12 miles, but also as regards other questions such as the adoption of the concept of the exclusive economic zone.

However, no agreement was reached and this Second Conference entailed no modification of the regime for the navigation of straits enshrined in the *Geneva Convention on Territorial Sea and Contiguous Zone* of 1958.

Only the tendency which can be seen in the extension of the width of the territorial sea involves a change, to the extent that the number of geographical straits which are considered to be international straits will increase substantially.

If we had to **synthesise** the legal situation of the straits addressed by the III Conference, we can say that *in peacetime* the majority of the straits in the world have been completely free, in such a way, that the type of right which exists in practice approximates more to passage on the high seas than to innocent passage. While *in wartime* a transit channel which is more equivalent to an innocent passage has always been available.[59]

We believe that it should be pointed out that, until this date, the width of territorial sea oscillated between 3 and 6 miles, which means that there were scarcely **ten international straits** included in territorial waters. Evidently, the regime of the 1958 Convention is applied to these, while a regime of freedom applies to the rest as there is a corridor to the high sea. A triviality if we consider that the existing straits amount to over 200.

Thus, the state practice we have referred to concerning navigation in the straits existing previously to the III Conference cannot be fully considered to be a consolidated norm of International law, while, to an even lesser extent, can it be applied to the new legal situation which will arise from this Conference. In fact, the

---

[56]Vid. Res. VIII, 27 April 1958, *UNCLOS II*, 1960, p. 145.

[57]A synoptic table of the widths of the territorial seas at this time can be consulted in Doc. A/CONF.19/4, in *UNCLOS II*, 1960, pp. 157–163.

[58]For a study of the proposals submitted, see Yturriaga Barberán (1991a, pp. 37–41).

[59]Cf. O'Connell (1984, p. 327).

enshrinement of 12 mile territorial waters by the UNCLOS will substantially change the legal scenario of straits, taking into account that this extension will involve an increase from 10 to more than 100 international straits whose waters fall within the territorial waters of one or several States.[60]

The position of the "new" coastal States with international straits will be underpinned by the defence of their sovereignty over these waters and, therefore, of the right of innocent passage codified in 1958, which will crash head on with the insurmountable wall of the Great Sea Powers which will apply pressure in order to achieve the freedom to navigate and fly over these straits.

[60]A map made by the United States Department of State raised to 116 the number of straits which would become international straits with the extension of territorial waters to 12 miles (cf. U.S. Dept. of State, Office of the Geographer, *World Straits Affected by a 12 Mile Territorial Sea*, Chart #510376, 1971).

Chapter 2
The Third Conference of the United Nations on the Law of the Sea or the Metamorphosis of Consensus

The situation enshrined in article 16.4 of the 1958 Convention failed to satisfy the Maritime Powers, suspicious of the implications of innocent passage in the straits which hindered the mobility of their hegemonic strategic deployment,[61] and also failed to satisfy a group of States interested in a number of important straits and, for ideological or conservationist reasons, these States wanted to impose a greater degree of control over the passage through the straits than the control over innocent passage.

Added to this unquestionable circumstance is the convergence of three factors[62] in the decade of the 1960s, which made the latent problems in the 1958 Geneva Convention concerning the right to air and sea navigation of the straits and contributed to the erosion of this right. Firstly, the technological development in ships and aircraft, the closing of the Suez Canal, and the growth of world trade has involved an increase in the size, nature and number of the ships and aircraft which use the straits. Secondly, the coastal States became preoccupied by the hazards of pollution and accidents in the straits.[63] Finally, there was a tendency to extend the territorial waters to 12 miles.

In this regard, inevitably a growing number of coastal States claimed 12 mile territorial waters and sustained that the passage of the straits included in their territorial waters was subject to the regime of innocent passage.[64] In addition, as

---

[61]Cf. Pastor Ridruejo (1983, p. 78).

[62]Cf. Grandison and Meyer (1975, p. 403). We must add the emergence of the so called countries of the Third World to these three factors, as these countries claimed a revision of the law in force which they had not participated in forming.

[63]In 1969, the collision and explosion of three super-tankers in the Santa Barbara Channel, the *Mactra*, the *Mapessa* and the *King Haakon VII*, led to the loss of the *Mapessa* and substantial damage to the other two vessels increased the preoccupation regarding the danger of pollution already started with the *Torre Canyon* incident in 1967.

[64]This was done by Iran in 1959, Yemen in 1967, Malaysia in 1969, Morocco in 1969, France in 1971, and Oman in 1972. A table showing the evolution of the claims concerning the different widths of territorial waters can be seen in Burke (1977, pp. 195–196).

A.G. López Martín, *International Straits*,
DOI 10.1007/978-3-642-12906-3_2, © Springer-Verlag Berlin Heidelberg 2010

a consequence of the fear of accidents and pollution, some of these States demanded the right to regulate the manner of passage and to establish the conditions of passage through the straits. Nevertheless, these claims were rejected by the main Maritime Powers.

Undoubtedly the revision of the Law of the Sea in general, and the Law concerning navigation through the straits in particular, was necessary. In this same respect, in the middle and at the end of the 1970s, there was a strong movement to promote this revision. Thus, in 1967, the USSR requested the opinion of the United States and other countries on the possible holding of a new inter-governmental conference on the Law of the Sea, in which a 12 mile limit for territorial waters would be established together with an exclusive fishing zone. The United States authorities responded that its position would be favourable if the USSR and other States were prepared to support provisions which recognised free passage through and above the international straits. After a number of consultations, in 1968, Soviet and United States experts drafted a project with three articles related to the establishment of 12 mile territorial waters, and the recognition of certain special fishing interests for the coastal States beyond the territorial waters, which was the subject of global distribution one year later.[65]

This growing pressure for the drafting of a new legal text to regulate the use of maritime areas appeared more specifically in the proposal put forward by the Maltese delegate, Mr Pardo,[66] before the United Nations on August 17, 1967, requesting the inclusion of the question concerning the use of sea and ocean beds in the agenda of the General Assembly;[67] a proposal which was extended on November 1, 1967.[68]

Finally, by Resolution 2367 (XXIII) of December 21, 1968, the General Assembly created a Special Committee with 42 members, which was later increased to 86 in 1970, and to 91 in 1971, and was responsible for preparing the conference on the Law of the Sea. Thus, the *Commission for Sea Beds* replaced the ILC as regards the work of codification and progressive development of International Law. The ILC had traditionally been in charge of this work in the area of the Law of the Sea, as we have seen in the previous paragraphs. The excessive workload of this organism and, consequently, the slow drafting of projects and the fact that numerous States considered that certain political matters were involved in the reform of the Law of the Sea, and this was in conflict with the ILC which is a legal organism, and led to the decision to create an *ad hoc* codifying organism for this occasion.

[65]Cf. Yturriaga Barberán (1991b, pp. 367–368).

[66]For a detailed analysis of the so called "Pardo Declaration" and its immediate consequences, see Marffy (1985, pp. 123–142).

[67]Cf. The title of the agenda proposed by Ambassador Pardo, in a *memorandum* explaining the "Declaration and treaty concerning the reservation exclusively for peaceful purposes of the sea-bed of the ocean floor, underlying the seas beyond the limits of present national jurisdiction, and the use of their resources in the interest of mankind" to the Secretary General (vid. Doc. A/6695).

[68]Doc. A/A1/PV 1515 and 1516.

Subsequently, through section (C) of Resolution 2750 (XXV) of December 11, 1970, the General Assembly decided to convene the III Conference of the United Nations on the Law of the Sea[69] which began in 1973.

The traditional conception was that the III Conference was established basically as a mechanism for the exploitation of the sea beds, as stated by B.P.V. Rao,[70] mistakenly. In fact, in the words of J.I. Charney, "the present Conference has its origin in the efforts of the United States and the Soviet Union in the early 1960s to protect their strategic interests in transiting the oceans, particularly international straits".[71] Such dissuasive strategic interests are of specific importance for the United States for the passage of its submerged nuclear submarines so that their detection in the straits will be impeded or hindered as their fundamental value lies precisely in their mobility and the difficulty to locate them; while the USSR has a geographical position which, in the majority of cases, forces its fleets to pass through straits in order to reach the high seas, which leaves it at a clear disadvantage with the United States. Therefore, a regime of innocent passage in the straits would have very serious consequences for its nuclear and conventional submarine forces.[72]

In this respect, the two substantial objectives of the main sea powers were undoubtedly focussed on refuting the claims for 12 mile territorial waters, and achieving a new right of passage in transit through the straits affected by 12 mile territorial waters. Therefore, it can be stated that these constitute some of the most crucial issues debated at the Conference.

This reached such a point that Resolution 2750 (XXV) expressly refers to both these substantial problems, and included the questions of the extension of territorial waters and international straits.

[69]A detailed presentation on the development of the work of the Commission of the Sea Beds and the III Conference of the United Nations on the Law of the Sea is in the Office for Ocean Affairs and the Law of the Sea (1992, in vol. I, pp. 21–152, there are details of the work of the Commission, while vol. II deals extensively with the III Conference). In addition, there is abundant scientific literature on such important work. Among the many works, we can cite some which refer especially to international straits: Burke (1977, pp. 195–220); Caminos (1987, pp. 66–120); Grandison and Myer (1975, pp. 393–450); Koh (1982, pp. 99–148); Maduro (1980, pp. 65–96); Marín López (1979 pp. 80–81, pp. 46–64); Momtaz (1974, pp. 841–859); Nordquist (1993, pp. 279–398); O'Connell (1982, pp. 328–331); Oxman (1985, pp. 143–216); Rao (1984, pp. 398–405); Robertson (1980, pp. 801–860); Stevenson and Oxman (1975, pp. 1–30); Yturriaga Barberán (1991b, pp. 366–396); Yturriaga Barberán (1991a, pp. 41–162). Also of interest is the excellent work of Jiménez Piernas (1982, pp. 815–931), which correctly analyses the question of the straits regarding waters in archipelagos, due to the unquestionable similarity of their legal regimes.

[70]Vid. Rao (1984, p. 403). Likewise Lawyer (1974, p. 1), for whom the matter of international straits "is probably the most important single issue to come before the conference".

[71]Vid. Charney (1977, p. 598).

[72]Cf. Jiménez Piernas (1982, pp. 807–808). The author adds the substantial strategic interests of France and the United Kingdom to those of the two main powers.

2.1 A Brief Summary of the Work Carried Out by the Commission on the Sea Beds as Regards International Straits

The period from 1971 to 1973 was marked by the preparation of the III Conference of the Commission on the Use of the Sea and Ocean Beds beyond the Limits of International Jurisdiction for Peaceful Purposes. Taking the previous resolution of the General Assembly as a mandate, the Commission divided its work among three sub-committees. Sub-committee I was in charge of preparing the project of articles on the international regime of sea beds beyond the limits of national jurisdictions. Sub-committee II was responsible for drafting a list of points for the conference agenda. Finally, Sub-committee III was assigned the issue of sea pollution and scientific research.

Throughout the 1971 session, the work of Sub-committee II was focused on the drafting of a complete list of questions concerning the Law of the Sea, including the regime of territorial waters and international straits, as well as the contiguous zones, the continental shelf, fishing and the high seas.[73] During the following two years, numerous statements and recommendations were put forward, and several projects of articles on each subject were presented. Undoubtedly, the question of navigation through international straits occupied a pre-eminent position in the preparatory work of the Commission from the beginning. The projects of articles presented could be grouped basically by interests in two categories: the developed naval powers and the developing coastal States of straits; the former tended towards freedom of navigation through the straits, while the latter were in favour of a regime of innocent passage.[74]

In short, the contradiction between both groups of States was reduced to two different conceptions as regards navigation through international straits, set out at three distinct levels: the legal nature of these waters, the naval powers did not accept that these were part of territorial waters, the navigation regime, and, as regards the distinction between commercial vessels and aircraft and warships and warplanes, this was irrelevant for the sea powers.

---

[73]The list drafted by Sub-committee II can be seen in *ILM*, vol. 11, 1972, p. 1174 et seq.

[74]The States in favour of the single regime of innocent passage were: Albania, Argentina, China, Egypt, Spain, Fiji, Iran, Kuwait, Morocco, Oman, Peru, Tanzania and Yemen. While Algeria, Bulgaria, Canada, Cuba, Czechoslovakia, Denmark, the United States, Finland, Ghana, Hungary, Iceland, India, Iraq, Israel, Italy, Liberia, Mongolia, Nigeria, Poland, the United Kingdom, the Democratic Republic of Germany, Singapore, Sri Lanka, Sweden, the Ukraine, and the USSR were to a greater or lesser extent in favour of the application of in transit passage in certain straits.

2.1.1 *The Proposals of the Main Sea Powers and the Freedom of Navigation*

As soon as the substantive work of the Commission began, the United States "opened fire"[75] with the submittal of a project of three articles which essentially reproduced the text agreed to in 1968 with the Soviet Union. While article I referred to the 12 mile territorial waters, and the third referred to fishing, article II claimed the freedom of the high seas for vessels and aircraft which crossed straits used for international navigation, in the following terms:

1. In straits used for international navigation between one part of the high seas and another part of the high seas or the territorial sea of another State, all ships and aircraft in transit shall enjoy the same freedom of navigation and over-flying aircraft, for the purpose of transit through and over such straits, as they have on the high seas. Coastal States may designate corridors suitable for transit by all ships and aircraft through and over such straits. In the case of straits where particular channels of navigation are customarily employed by ships in transit, the corridors, so far as ships are concerned, shall include such channels.
2. The provisions of this article shall not affect conventions or other international agreements already in force specifically relating to particular straits.[76]

The first two articles of the United States project constituted an indivisible package. In the words of Stevenson[77] before the Commission, the United States would only accept 12 miles territorial waters in exchange for the recognition of freedom of navigation through and over international straits.

This proposal received the total support of the Soviet Union and the United Kingdom; other countries, such as Australia, the Netherlands and Norway were also in favour. However, several States considered that the regime of innocent passage was adequate and did not see the need for the proposal of the United States; this was the case of Denmark, Greece and Italy. Other States such as Spain were more preoccupied by the need to establish control over pollution considering the hazards of super-tankers. In the light of these observations, the representative of the United States made another declaration during the 1972 session of the Commission, and repeated the promise that, in accordance with the 1971 project, the rights of vessels and aircraft in transit were not equivalent to the freedom of navigation on the high seas. In order to ensure that vessels and aircraft would not contaminate or navigate, Stevenson proposed that the coastal States adopt regulatory systems compatible with the norms established by the international organisms – the then International Maritime Consultative Organization (IMCO) and the International Civil Aviation Organization (ICAO).[78]

---

[75]Cf. Yturriaga Barberán (1991b, p. 369).

[76]Vid. The text of the project of the United States in Office for Ocean Affairs and the Law of the Sea I (1992, p. 27). For a detailed analysis of this project, which supposes a repetition of the Declaration of President Nixon on May 23, 1970 on sea policy, see Knight (1972, pp. 759–788).

[77]Cf. Robertson (1980, p. 808).

[78]Cf. UN Doc. A/AC.138/SC.II/SR.33-47 (1972).

During the 1972 session, the USSR submitted a project similar to the one submitted by the United States although it applied exclusively to the straits which join two parts of the high seas. However, this proposal contained a number of provisions on the obligations affecting vessels and aircraft in transit and included a clause on compensation for damages caused to coastal States as a consequence of over-flying aircraft;[79] although it made no mention of the obligation to comply with neither the rules of the coastal State nor the power of this State to issue norms on pollution. Subsequently, in the 1973 session, the Soviet representative, Mr Malik, pointed out that the USSR was prepared to extend the regime stipulated in its proposals to the straits which join the territorial waters of a State with the high seas.[80] Numerous delegations were concerned by this coincidence of opinions of the two main naval powers.

It was clear that a interesting change had occurred in the position of the Soviet Union as concerns the passage of warships; while article 23 of the 1958 Geneva Convention was the subject of a Soviet reservation, in the sense that warships could not cross the territorial waters of another State, including the waters of straits, without the previous authorisation of the coastal State,[81] in its 1972 proposal the USSR made no distinction between warships and merchant ships.

Despite the efforts of the United States and the Soviet Union to present their projects as balanced positions on the question of straits, many of the coastal States presented an unshakeable position involving only the recognition of innocent passage through the straits.

2.1.2 The Propositions of the Coastal States and the Right of Innocent Passage

Throughout the last two sessions of the Sea-Bed Commission held in 1973, the United States and the USSR were challenged by numerous proposals from coastal States in straits,[82] which trusted that these might serve as a draft text for the III Conference in December 1973.

Spain was the first country to react against the United States and Soviet proposals, and pointed out that, due to a generally accepted traditional rule, vessels have the right of innocent passage which cannot be suspended and, according to the stipulations of the 1944 Chicago Convention, civil aircraft have freedom of transit and stopover in the territory of other States for non-commercial purposes, while the transit

[79] Cf. UN Doc. A/AC.138/SC.II/L.7 (1972).

[80] Cf. UN Doc. A/AC.138/SC.II/SR.69 (1973).

[81] Cf. http://untreay.un.org.

[82] The total number of proposals submitted to the II Sub-Committee in 1973 amounted to approximately 50, a figure which clashed with the two proposals in the 1971 and 1972 sessions (cf. Caminos 1987, p. 73).

of non-civil aircraft requires the previous authorisation of the State being over flown. The Spanish representative, Mr Sr. Ruiz Morales, concluded b saying that:

> if these intended liberties concerning navigation and over flying aircraft in international straits are enshrined in law, the final result will be to establish a right of indiscriminate transit through straits for the benefit of a few ... and this will be directly in favour not of civil navigation but of military aircraft, which are at present not allowed passage, and warships, especially submarines.[83]

Similar declarations were made by Denmark, Indonesia, Malaysia, Ethiopia and Italy. The Spanish delegation also handed out a "Memorandum on the Question of International Straits", in which it expressed its opposition.[84] In addition, it encouraged the constitution of a *Group of Coastal States of Straits* in favour of the regime of innocent passage. This group, made up of Cyprus, Spain, the Philippines, Greece, Indonesia, Malaysia, Morocco and Yemen, submitted a project of articles on "Navigation in territorial waters, including the straits used for international navigation",[85] at the session of the Sea-Bed Commission held from March 3 to April 6, 1973. The proposal included 23 articles, and its preamble listed five basic objectives which had to be taken into account in the draft version:

1. Navigation through the territorial sea and through straits used for international navigation should be dealt with as an entity since the straits in question are or form part of territorial seas;
2. Regulation of navigation should establish a satisfactory balance between the particular interests of coastal States and the general interest of international maritime navigation. This is best achieved through the principle of innocent passage which is the basis of the traditional regime for navigation through the territorial sea.
3. The regulation should contribute both to the security of coastal States and to the safety of international maritime navigation. This can be achieved by the reasonable and adequate exercise by the coastal State of its right to regulate navigation through its territorial sea, since the purpose of the regulation is not to prevent or hamper passage but to facilitate it without causing any adverse effects to the coastal State.
4. The regulation should take due account of the economic realities and scientific and technological developments which have occurred in recent years; this requires the adoption of appropriate rules to regulate navigation of certain ships with 'special characteristics'.
5. The regulation should, finally, meet the deficiencies of the 1958 Geneva Convention, especially those concerning the passage of warships through the territorial sea, including straits.

This project adopts a monistic approach to the regime of straits, following the directives of the 1958 Geneva Convention on the suspension of innocent passage although it did include important innovations as regards other aspects. For example,

---

[83]Vid. The first intervention of the Spanish representative in Office for Ocean Affairs and the Law of the Sea I (1992, pp. 30–32).

[84]The four arguments on which Spain grounded its position were as follows: strategic reasons, the satisfactory functioning of the regime of innocent passage, easements of the coastal States of straits and the false accommodation of interests.

[85]Cf. UN Doc. A/AC.138/SC.II/L.18 (1973).

it tried to objectively specify the concept of innocent passage as it indicated the activities which the vessels were prohibited from carrying out during passage through the straits, such as espionage, propaganda and interference with the communications systems, illicit trading, exploration or exploitation of resources, the embarking or disembarking of persons or things. It introduced a non-discrimination clause as regards the flag of the vessel, the nationality of the passengers, the place of departure and the destination. Furthermore, it reinforced the competence of the coastal States to control navigation with regard to security and maritime traffic, the placement and use of installations to assist navigation and the exploitation or exploration of marine resources, maritime transport, research into the marine environment, and the passage of vessels with special characteristics.

However, the project contained no provisions concerning over flying aircraft as it considered that the rules of the Chicago Convention continued to apply.

Similar to this proposal by the eight "Strait" States was the one submitted by Fiji.[86] We should point out that although the position of Fiji on commencement of the preparatory work seemed to be aligned with those States opposed to an autonomous regime concerning international straits, its position changed dramatically throughout the first two sessions of the III Conference, when, together with the United Kingdom, it led a private negotiation group regarding the straits, which drafted what would later become the articles finally included in the UNCLOS.

The original and premonitory proposals submitted by Malta had little effect on the work of the Sea Bed Commission as regards its global and integrated conception of the maritime areas. In 1971 it presented an ambitious treaty project on marine areas[87] which, to a certain extent, followed the regime of innocent passage, with the only difference that it recognised free passage in straits over 12 miles wide and granted the right of innocent passage to aircraft. It also proposed that the right of the coastal State to prevent the passage of foreign vessels only went as far as the first 6 miles from the coast. In 1973, it submitted a corrected, extended version of its proposal,[88] in which certain special competences were recognised for the coastal States within a 12 mile band adjacent to the coast. These competences, in the case of the straits whose width was less than 24 miles, were restricted to the possibility of demanding the following: that the vessels comply with the obligatory systems for the separation of traffic, their passage be uninterrupted and fast, that they use pilots, that warships notify passage three days in advance and that unidentified submarines navigate on the surface. It recognised the right of aircraft to fly over the straits, and this could only be suspended by the coastal State in the event of grounded fear of serious and imminent threat to its security.

However, the project which had most effect was the very brief one submitted by Italy, which stipulated a dual regime for straits:[89] 'innocent passage' for those straits

[86]Cf. UN Doc. A/AC.138/SC.II/L.42 (1973).

[87]Cf. UN Doc. A/AC.138/53 (1971).

[88]Cf. UN Doc. A/AC.138/SC.II/L.28 (1973).

[89]Cf. UN Doc. A/AC.138/SC.II/L.30 (1973).

with a width less than 6 miles, whose coasts belonged to one State and which had other alternative nearby navigation routes; and 'free transit' for the other straits, where vessels and aircraft would have the same freedom to navigate and fly over as on the high seas, as regards transit through or over straits which communicate two parts of the high seas or apart of the high seas with the territorial waters of a foreign State.

In the words of J.A. Yturriaga Barberán, "this proposal involved a high level of cynicism and opportunism as it was drawn up tailored to the Strait of *Messina*, which remained safeguarded by the often abused regime of innocent passage, while the other straits came under the regime of free transit".[90] This project was fiercely criticised by the Spanish representative.

It is evident that the projects and proposals submitted to the Commission throughout these three years are a result of two approaches to the regime regulating navigation through international straits which are diametrically opposed. On the one hand, there is the standpoint of the naval powers based on an autonomous regime of freedom of air and sea navigation. On the other hand, there is the position of the coastal States of straits as regards maintaining a unitary regime grounded on the right of innocent passage.

The variety and disparity of the proposals made it impossible to compile all of them in a single document, while this also made it unfeasible for the Sea Bed Commission to draw up a draft of articles to be considered at the III Conference of the United Nations on the Law of the Sea. In fact, the final report adopted on August 23, 1973 is only a list of nine possible variants of how the articles concerning item 4 (straits used for international navigation) could be drafted.[91]

This peculiar circumstance undoubtedly supposed a change of the methodology used at other codification conferences held under the auspices of the United Nations.

2.2 The Most Relevant Aspects of the Development of the Question of the International Straits at the III United Nations Conference on the Law of the Sea

On December 3, 1973, the III United Nations Conference on the Law of the Sea began its long, complex passage in *New York* although this first period of sessions was dedicated exclusively to organisational questions. The first substantial discussions took place during the *second period of sessions* held in *Caracas* from June 20 to August 29, 1974. Although the first session of authentic negotiation was the third one, held in Geneva in 1975, where a *Single Informal Negotiation Text* was achieved and this would give fundamental shape to the new Convention. The dilemma involving innocent passage and freedom of navigation – or, in other words, Straits

[90]Vid. Yturriaga Barberán (1993, p. 376).

[91]The explanation of the variants adopted in the Report of August 23, 1973 can be consulted in the Office for Ocean Affairs and the Law of the Sea I (1992, p. 138 et seq).

States and Maritime Powers – focused the debates on the regime of straits. However, the initial resistance of the coastal States was gradually weakened during the successive periods of sessions faced with the strong pressure of the Superpowers and the great powers, as well as the neutrality or passivity of the immense majority of the Group of 77, and lost force and consistency. In fact, only Spain showed its disagreement concerning some of the provisions of the new regime up to the last moment and undoubtedly was one of the more active States against the great powers.

2.2.1 Formulation of the Principal Tendencies

At the beginning of the second period of sessions in Caracas in 1974, the functioning of the Conference was organised, by being structured in accordance with the model of the Sea-Bed Commission: a First Committee in charge of the international regime and the mechanism for the regulation of the use of the sea beds beyond the limits of national jurisdiction; a Second Committee dealt with all the traditional matters concerning the Law of the Sea, including territorial waters, the economic zone, the continental shelf, the high seas, the straits, the archipelagos and fishing; and Third Committee attended to the problems of sea pollution, conservation, scientific research and technological transfer. This session turned into a forum for presenting the various political declarations together rather than a negotiation session; in fact, the Second Committee only achieved the preparation of a document entitled "Working Paper of the Second Committee: Main Trends",[92] a collection of alternative formulations amounting to 127 pages. This was difficult to deal with and constituted a compendium of the various positions rather than a negotiation text, which was in accord with the general content of the session.

As regards international straits, a total of nine proposals concerning or referring to these were submitted; the basic positions presented by the States to the Sea-Bed Commission were again explained, proposals were repeated and some new ones were submitted.

Thus, the United States insisted with its already known position that the 12 mile extension of the territorial sea should suppose the recognition of a non-discriminatory right to free transit through international straits as regards air, surface and submarine navigation.[93] The USSR cosponsored a project similar to the one submitted in 1972 together with another five socialist countries.[94]

Spain continued to oppose the proposals for free transit, and presided and coordinated the work of the Group of Coastal States which drafted a version of

[92]Cf. Doc. A/CONF.62/C.2/WP.1.

[93]Vid. The Declaration of Ambassador Stevenson in this respect in *Documentos Oficiales de la Tercera Conferencia de Naciones Unidas sobre Derecho del Mar*, New York, 1975, vol. I, p. 179.

[94]The project was cosponsored by Bulgaria, Czechoslovakia, Poland, the GDR, the Ukraine and the USSR (cf. Doc. A/CONF.62/C.2/L.11, 1974).

the 1973 text in which, although some minor concessions were made concerning maritime navigation, silence ensued as regards flights over straits. For tactical reasons, the new text was submitted by Oman although it was subsequently cosponsored by Malaysia, Morocco and Yemen.[95] In addition, Fiji presented a revised version to its proposal of 1973.[96]

Another two propositions referred exclusively to the definition of straits used for international navigation; one was presented by Canada,[97] the other was cosponsored by ten Arab States (Algeria, Bahrain, Iraq, Kuwait, Libya, Qatar, Saudi Arabia, Syria, Tunisia and the United Arab Emirates).[98] The Dominican Republic also made a proposal, establishing that the principle of previous notification should be adopted by the States which had bays and straits in common, before any of them carried out work or placed installations which might generate pollution of any kind for the other State.[99]

However, undoubtedly, the most important proposal of all those presented at this second session in Caracas was the one put forward by the United Kingdom, which assumed a supposedly mediating function as it was both a maritime power and a coastal State of one of the most important navigable straits in the world, the Dover Strait, and had to seek the reconciliation of all the interests involved. The project was submitted as an attempt to find a balance between the proposals for freedom of navigation and the traditional support for innocent passage.

As the regime of 'free transit' had received severe criticism, the United Kingdom very skilfully presented a proposal which introduced the concept of 'passage in transit' through the straits used for international navigation which communicate two parts of the high seas; this passage consisted of the exercise of freedom of navigation and over-fling planes exclusively for the purposes of continuous and rapid transit through the strait. The introduction of the new term 'passage in transit' had the advantage that it avoided the excesses of the previous proposals. The seven key factors of this project are as follows:[100]

1. The 'passage in transit' is only applied to straits used for international navigation which communicate two parts of the high seas; other straits are governed by the right of innocent passage which cannot be suspended. The straits formed by an island and the mainland are excluded from passage in transit if there is a similar

---

[95]Cf. Doc. A/CONF.62/C.2/L.16 (1974).

[96]Cf. Doc. A/CONF.62/C.2/L.19 (1974). Algeria also submitted a proposal concerning navigation through straits which communicated two parts of the high seas and served as access to semi-enclosed sea; in such straits, the merchant ships and the ships of a State exploited for commercial purposes had freedom of transit, while the warships and ships of a State which were engaged in non-trading activities had innocent passage. No mention was made of over-flying aircraft (cf. Doc. A/CONF.62/C.2/L.20, 1974).

[97]Cf. Doc. A/CONF.62/C.2/L.83 (1974).

[98]Cf. Doc. A/CONF.62/C.2/L.44 (1974).

[99]Cf. Doc. A/CONF.62/C.2/L.59 (1974).

[100]Cf. Doc. A/CONF.62/C.2/L.3 (1974).

navigation route on the other side of the island. The straits which have a high seas route are also exempt.
2. While in passage in transit, the ships and aircraft must navigate without delay and must not carry out other activities than those which are normal in transit. The vessels must comply with the generally accepted international norms on safety at sea and the prevention of pollution. The aircraft must observe the rules of the ICAO. The aircraft of a State are normally bound by these rules and must act with due diligence as regards safety.
3. The coastal States of straits have the power to adopt schemes for the separation of traffic and sea routes and have these complied with once they are approved by the proper international authority.
4. The coastal States of straits have the power to adopt norms which put the international regulations on the prevention of pollution into effect and have these complied with.
5. The coastal States of straits cannot hinder passage and must notify the hazards to sea navigation and overflying aircraft.
6. In the case of straits governed by the right of innocent passage which cannot be suspended, the norms must specify which acts are considered to be innocent.
7. The State of the flag of public ships and aircraft must be liable for the damage caused as a consequence of failure to comply with the laws of the Coastal State of straits.

The three proposals which were the subjects of most of the discussion during the second period of sessions of the Second Committee were those of the Socialist States, the proposal of Oman and the proposal of the United Kingdom, although it was this last proposal that dominated the debates fundamentally due to two reasons. On the one hand, because it was considered to be a compromise formula and received the support of the majority of the delegations; on the other hand, because even the delegations which did not fully support the proposal acknowledged the tendency to support it.

As we have mentioned, this second period of sessions concluded with no agreement on the projected articles. However, the proposals submitted were collected in a document on the formulation of the main tendencies drafted in a way which was not very objective as was shown by the fact that, in the section concerning ‘innocent passage’, there were five variants of the defence of free transit, and the regime of innocent passage was maintained as rather insignificant for certain unimportant straits; or also, that the position of the defenders of innocent passage as a general regime only appears in a footnote.

2.2.2 *Preparation and Drafting of the Negotiation Texts*

After the second period of sessions, the work of the Second Committee was carried out mainly at informal meetings, informal consultative groups, and work groups.

Neither was progress in the negotiations recorded in the summary reports of the formal meetings of the Committee, which made little progress.

In this regard, during the *third period of sessions*, held in *Geneva* from March 26 to May 10, 1975, a special group was constituted to deal with straits within the Second Committee, where – in a dialogue of the deaf[101] – the theses of the supporters of innocent passage again confronted those supporters of free transit. However, once the Conference entrusted the Presidents of the Committees with the drafting of extra-official texts, the negotiation was transferred to other forums, specifically, those of the main interest groups which were acting parallel to each other with no attempt at negotiations with each other. On the one hand, there was the "Group of Coastal States of Straits", and on the other, the "Private Group on Straits", an informal group co-presided by Great Britain and Fiji, and made up of Argentina, Australia, Bahrain, Bulgaria, Denmark, the United Arab Emirates, India, Iraq, Iceland, Italy, Kenya, Liberia, Singapore and Venezuela.[102]

In the session held on May 1, 1975, Oman submitted a document adopted by consensus in a private group, and this insisted on three points:[103]

1. The presumption of innocent passage in the straits used for international navigation was sufficient.
2. The basic point was to avoid collisions.
3. Overflying aircraft must be regulated by the Chicago Convention, not by a treaty on the Law of the Sea.

The Private Group on Straits drew up revised version of the British proposal of 1974, which the Group called a consensus text, as it specified that its objective was an adaptation of the proposals of Fiji and of the United Kingdom in order to achieve a balance between the interests of the coastal States of straits and the maritime powers. The main innovations of this project were as follows:[104]

1. The assertion that the activities which did not entail the exercise of the right of passage in transit would be regulated by the relevant norms of the Convention.
2. The addition of two new requirements for the passage of ships and aircraft.
3. Precisions on the establishment of maritime routes.
4. The extension of the regulatory competence of the coastal State: security of navigation and the regulation of maritime traffic, the prohibition to fish and the prohibition to embark or disembark products, currency or persons.
5. The assertion that the regime on transit would not affect the legal condition of the waters of the straits.

[101]Cf. *Ibidem*, p. 383.

[102]With the exception of Fiji, all the States which composed this Group had submitted proposals in favour of free transit or had supported these. Fiji was won over to the 'cause' by the United Kingdom.

[103]Cf. Caminos (1987, p. 97).

[104]Cf. Yturriaga Barberán (1993, pp. 235–236).

This project was the subject of harsh criticism by Canada, Chile and Norway, which considered that its definition of straits included a number of straits which they considered to be internal waters. They also submitted an official memorandum with their viewpoints to the President of the Second Committee.[105]

On the final day of the Geneva session, President Amerasinghe announced the distribution of the *Informal Single Negotiating Text*, accompanied by a note from the President which stated that this text "would serve as a procedural device and only provide a basis for negotiation".[106]

The tactic of waiting until the last minute to make the *Informal Single Negotiating Text* public was a success to the extent that it prevented the delegates from checking the new text and reopening the debate. Part II of the *Informal Single Negotiating Text* contained the articles project submitted by the Second Committee, which included a section on navigation through international straits,[107] which accepted the thesis of dual regimes in territorial waters and straits and this entailed the triumph of the naval powers and the defeat of the coastal States of straits.[108]

In fact, the provisions which made up this section were a reproduction of the proposal prepared by the Private Group of States cosponsored by the United Kingdom and Fiji, with some modifications to the definition of straits introduced in order to satisfy the claims submitted by Canada, Chile and Norway, together with certain additions, such as the inclusion of a clause which stipulated that all activities which do not entail the exercise of the right of passage in transit would be subject to the other applicable provisions of the Convention, and the insertion of two new obligations for ships and aircraft in transit: to abstain from all activities not related to the normal modalities of transit and to comply with the relevant provisions of the Convention.

This strategy involving the inclusion of proposals with a wider acceptance sought to ensure that the text prepared was accepted by the majority of the segments of the Conference. At the time, what it did achieve was to discourage the Group of Coastal States of Straits, and even led to a crisis within this Group which stopped meeting and ended up being dissolved. From this time, each of its members recuperated their freedom of action which allowed them to act separately in defence of their own interests.

Thus, the *Informal Single Negotiating Text* confirmed the diversity of regimes within straits, as well as the acknowledgement of an autonomous legal regime for straits, which meant a break with the position codified in the 1958 Geneva Convention.

The successive versions of the Negotiation Text did not involve important revisions of the articles, which mean that the efforts made until this date had

[105]The text of this memorandum can be seen in Platzöder (1983, pp. 223–224).

[106]Vid. Doc. A/CONF.62/SR.54, p. 3.

[107]Cf. Doc. A/CONF.62/WP.8 (1975).

[108]This defeat also involved the archipelago States, as it ignored their proposals on the navigation regime in the waters of archipelagos, which explains the coincidence of the navigation regimes.

given their fruits. Specifically, as regards the provisions related to passage through the straits used for international navigation, these appear in the successive negotiation texts with no substantial changes. In fact, despite the efforts of a small minority, which basically included Spain, Greece, Malaysia and Morocco, to modify the articles related to passage in transit in 1976, 1977 and 1978, none of the amendments submitted received substantial support.

Along these same lines, the provisions of Part II of the *Informal Single Negotiating Text* were again included in the *Revised Single Negotiating Text*, given out on the last day of the eighth week of negotiation in *the fourth period of sessions*, held in New York in May 1976, as "a basis for continued negotiation".[109] Due to the informal nature of the *Informal Single Negotiating Text*, which could not have any official amendments, some former members of the Group of Coastal States of straits submitted extra official amendments. Thus, Greece submitted a number of amendments related to 'overflying aircraft'.[110] Spain was open to the regime of passage in transit for maritime navigation but continued to oppose the application of this regime to overflying aircraft; in this regard, it proposed the suppression of any reference in the text to aircraft or overflying, it tried to equate the regime on straits with territorial waters as far as this was possible as regards the rights and obligations of the coastal State and the vessels in transit, it maintained the obligation for submarines to navigate on the surface, and included a clause on the responsibility of vessels not covered by immunity.[111]

However, after numerous discussions held by the groups present, the modifications to the *Un-official Single Text for the Purposes of Negotiation* included in the *Revised Single Negotiating Text* were more formal than in substance. The minor changes made included: the replacement of the expression "Straits States" by "Coastal States of Straits", which entailed the omission of the definition of "Straits State" contained in paragraph 3 of article 34; the term "legal status of the straits" in article 35(c) was replaced by "legal regime of the straits"; the terms "with similar hydrographical and navigation characteristics" were added to articles 36 and 38.1; on the suggestion of the Guyana delegation, the word "sovereignty" was added to paragraph (b) of article 39; moreover, the articles on the lateral passage to enter a coastal State of a strait and innocent passage were redrafted.

Three months later, the *fifth period of sessions* of the Conference took place, and there were no modifications made to the Single Revised Negotiation Text, although the subject of the straits used for international navigation continued to appear in the list drawn up by the President of basic subjects pending solution and which the Committees should focus their attention on.[112]

In 1977, again in New York, the *sixth period of sessions* was held and focused mainly on the question of the exploitation of the sea beds. The provisions on this

---

[109]Cf. Doc. A/CONF.62/WP.8/Rev.1 (1976).

[110]Cf. Platzöder (1983, p. 282).

[111]Cf. *Ibidem*, pp. 274–275.

[112]Cf. Doc. A/CONF.62/L.12/rev.1 (1976).

question were the only ones which were modified in the turn. The new text presented by president Amerasinghe is known as the *Informal Composite Negotiating Text*,[113] which combines the work of the three Principal Committees in a single negotiation text, and is the result of the two preceding texts. Its nature is defined by the President of the Conference in the following terms:

> Il ne s'agira absolument pas d'une texte de projet fondamental, comparable à celui que la Commission du Droit International avait établi à l'intention de la Conférence de Genève de 1958, et dont toute disposition qui n'aurait pas été rejetée à la majorité requise serait conservée.[114]

This is a text which will serve simply as an instrument of work and the basis of negotiation.

The document is structured into 16 parts, with seven annexes, and a total of 303 articles numbered consecutively. Despite the multiple extra official amendments presented,[115] the section on the passage through the straits remained basically unaltered, except for the inclusion of a new article prohibiting the ships from carrying out research activities during their passage through the straits (article 40), and a modification which was inserted into article 234, recognising limited powers of execution to the coastal State when ships infringed its provisions concerning the security of navigation or the prevention of pollution. This modification was the result of the negotiations made by a group of States directly interested in the consequences of the provisions on safeguards in the straits.[116]

The constant presentation of amendments by some and their being ignored buy others is also a reflection of the attitude of a few States that negotiation on straits had not been finalised, and of the position of the President of the Committee who considered that the provisions concerning straits were justly established and were a reflection of the tendency of the Conference.

During the first part of the *seventh period of sessions* held in Geneva in 1978, Spain handed out a "Memorandum on the question of aircraft flying over the straits used for international navigation" among the participating delegations, and this criticised the provisions of the *Informal Composite Negotiating Text* as regards aircraft flying over straits and submitted extra official suggestions which were never discussed or negotiated. It also presented a number of amendments in which, although concessions were made as regards maritime navigation, it continued to

---

[113]Cf. Doc. A/CONF.62/WP.10 (1977).

[114]Vid. Doc. A/CONF.62/L.20 (1977).

[115]Throughout this sixth session, Malaysia, Spain and Morocco presented several informal amendments which tended to specify the acts vessels in transit were forbidden to carry out, to extend the regulatory powers of the coastal State and to include a provision on objective liability foor damages (cf. Platzöder 1983, p. 397, pp. 394–395, and pp. 399–400, respectively for the amendments submitted by each State).

[116]This was explained in this way by President Amerasinghe (cf. Doc. A/CONF.62/WP.10/Add.1, 1977). In fact, the inclusion of the new article 40 and the amendment to article 234 were the result of bilateral negotiations held by Malaysia, Indonesia and Singapore, on the one hand, and by the United States, the USSR, the United Kingdom and Japan, on the other.

oppose any reference to flying over straits. Greece repeated its amendments of 1976, while Morocco and Yugoslavia repeated their amendments of 1977. None of these were accepted. In fact, throughout the subsequent, successive revisions which the *Informal Composite Negotiating Text* made at the Conference – in 1980 and 1981 – until the Convention Project was achieved, none of the amendments presented on the regime of transit through straits used for international navigation were included, except for the introduction of a general clause on liability for damages (article 304), as had been repeatedly claimed by Spain and Morocco.

Throughout the first part of the *11th period of sessions*, which took place in New York in 1981, it was evident that it would not be possible to adopt the text of the Convention Project by consensus, therefore, the President of the Conference opened the period for the presentation of amendments. Greece and Spain seized the opportunity to try to restrict planes flying over straits by several amendments, which were withdrawn a year later due to the pressure put on the amending delegations by President Koh. Only two Spanish amendments were maintained.[117] One regarding article 39 which proposed the suppression of the term "normally", was submitted to a vote and rejected by 55 votes against, 21 in favour and 60 abstentions. Another amendment to article 42 relative to the substitution of the word "applicable" by "generally accepted"; with 62 votes in favour, 29 against and 51 abstentions. This amendment was not approved as it did not receive the favourable votes of the majority of the States participating in the Conference, as stipulated in article 39.1 of the Regulations.[118]

The Voting on the Convention Project as a whole took place on April 30, 1982, during the second part of the 11th period of sessions. The text was adopted by 130 votes in favour, 4 against (the United States, Israel, Turkey and Venezuela), and 17 abstentions (including Spain). The formal signing of the Final Minutes of the Conference and the *United Nations Convention on the Law of the Sea* took place in Montego Bay (Jamaica), during the third part of the 11th session, on December 10, 1982. It came into force on November 16, 1994, once 12 months had elapsed since the deposit of the 60th instrument of ratification by Guyana. It currently has 155 parties.

Despite initial reticence, Spain signed the Final Minutes of the Conference on December 10, 1982 and signed the United Nations Convention on the Law of the Sea on December 4, 1984 (five days before the period established by article 305.2 of the Convention), ratified it on January 15, 1997.[119] Both the Spanish signing and the ratification instrument were accompanied by a number of interpretative declarations, using a power permitted by article 310 of the UNCLOS, most of which concerned the issue of international straits.[120] However we should point out that, at

[117]Cf. Doc. A/CONF.62/L.109 (1982).

[118]Vid. The text of the Regulation in Doc. A/CONF.62/30/Rev.2, New York, 1976.

[119]Vid. *Official State Gazette (BOE)*, No. 38, of February 14, 1997.

[120]Specifically Spain made five declarations on straits at the time of signing, and specifically qualified articles 39, 42, 221 and 233 of the UNCLOS.

the time of the ratification, Spain carried out a reordering and reformulation of the declarations related to the regime on navigation through straits.

Thus, the two declarations related to the protection of the marine environment, specifically, number 4 concerning article 42.1(b) and 7 on the interpretation of article 233 in accordance with article 34 were suppressed. The second declaration was maintained although it was redrafted in order to eliminate the references to the regulation of air space and given a wider meaning to include the content of the two suppressed declarations.[121] Finally, the other two declarations on the regime of international straits remained intact, and this was advisable, in the opinion of C. Jiménez Piernas, "in order to avoid interpretations by the user States of these straits which are clearly contrary to the legitimate interests of the coastal States in order to avoid the misuse of the passage in transit of aircraft and to safeguard their right to intervene in the event of accidents".[122] The remaining interpretative declarations made on signing were maintained substantially, at the time of ratification, these were those concerning the contentious issue of Gibraltar (2nd), which was correctly described by the same author as "playing to the gallery",[123] the fishing regime in the European Economic Area (4th), and the regime on the exploitation of the area (5th). Another two, whose justification was self evident were added to these, the 1st concerning the condition of a Member State of the EU, and the 6th in accordance with article 287, whereby Spain chose the ICJ in order to solve controversies related to the interpretation or application of the UNCLOS.

As pointed out by R. Riquelme Cortado, the basis of the Spanish declarations on the new regime of passage in transit through the straits used for international navigation lies "in the fact that the sovereignty of the State over the waters of the straits, its sea bed, its subsoil and the air space above must be reflected in the powers and competencies recognised to the coastal State as regards the regulation of air traffic and the prevention and control of pollution in this international route".[124]

Finally, it can be concluded that, except for a few insubstantial changes made by the Preparatory Committee, the text of the articles related to the straits included in

---

[121]The final drafting of the declaration establishes:

> 3. a) The regime established in Part III of the Convention is compatible with the right of the coastal State to dictate and apply its own regulations in the straits used for international navigation on condition that it does not hamper the right of passage in transit.

[122]Vid. Jiménez Piernas (2001, p. 113).
The meaning of these two declarations is as follows:

> b) In article 39, paragraph 3 a), the word 'normally' means 'except in cases of force majeure or serious difficulties'.
> c) The stipulations in article 221 does not deprive the coastal State of straits used for international navigation of the competences recognised by International Law regarding intervention in the case of accidents referred to in the article cited.

[123]Vid. Jiménez Piernas (2001, p. 113).

[124]Vid. Riquelme Cortado (1990, p. 185).

the *Informal Composite Negotiating Text* is identical to the one appearing in Part III of the final text of the *United Nations Convention on the Law of the Sea*, which finally leads us to the proposal of a the Private Group of States cosponsored by the United Kingdom and Fiji.

Considering this state of affairs, there is no doubt that, with regard to the legal regulation of navigation through international straits, this give and take concerning the 'package deal' at the III Conference lead to what could call the "desubstantiation of the agreed rule" and others call the "exhaustion of consensus".[125] Al last, the success of the Great Powers.

As was stressed by B.D. Smith:

> the maritime States at the Conference sought and achieved a straits regime of diminished coastal authority far more protective of military and economic navigational interests than the rules of innocent passage. The capacity of the coastal State to interfere with navigation in exercise of jurisdiction over environmental matters was specifically targeted for and subjected to reduction. In short, **the coastal State's jurisdiction over the condition and character of vessels in straits' transit is non-existent; jurisdiction over affirmative conduct affecting the environment is much diminished.**[126]

[125] Cf. Wengler (1974, p. 337).

[126] Vid. Smith (1988, pp. 208–209) (the bold lettering is ours).

Chapter 3
The Definition of the 'Undefined' Straits Used for International Navigation

It is true that the ocean is subdivided into marginal basins and seas, all of which are interconnected by straits.[127] From the purely geographical point of view, straits are considered to be "a narrow passage between two large bodies of water which have different density stratification and sea level elevations due to the independent, natural processes in each basin".[128] In this regard, the International Hydrographical Organisation defined geographical straits as "a narrow passage between landmasses or islands or groups of islands connecting two larger sea areas".[129] Also from a geographical viewpoint, although more concise, is the concept provided by G. Gidel, who stated that straits are "un passage maritime resserré entre deux terres".[130]

However, these purely geographical definitions, although they serve as a reference, they are not valid as regards providing a legal concept for this maritime area from the point of view of Public International Law. Taking into account that not all the geographical straits come within the area of application of the specific regime stipulated in Part III of the 1982 Convention, but only those straits used for international navigation, as stated in this third Part

As pointed out by Hyde[131] it is evident that 'there are straits and straits'. And only the straits "used for international navigation" are classified as 'international straits', and only these straits fall within the specific regime stipulated in Part III, sections 2 and 3, of the 1982 Convention.[132]

In fact, Part III of the UNCLOS is entitled: "Straits used for International Navigation". The use of these terms shows that the 1982 Convention clearly follows

---

[127]Cf. Pratt (1990, p. xi).

[128]Vid. *Ibidem*, p. 373.

[129]Vid. International Hydrographic Organization (2006, p. 94).

[130]Countering this, this author states that "est donc juridiquement un détroit tout passage naturel entre deux côtes, n'excédant pas de une certaine largeur et faisant communiquer entre elles deux parties des espaces maritimes" (vid. Gidel 1934, pp. 729–730).

[131]Cf. Hyde (1945, p. 487).

[132]Cf. Office for Ocean Affairs and the Law of the Sea (1989, p. 63).

A.G. López Martín, *International Straits*,
DOI 10.1007/978-3-642-12906-3_3, © Springer-Verlag Berlin Heidelberg 2010

in the wake of the 1958 Geneva Convention which conventionally confirmed this expression following the dictates of the ICJ in the *Corfu Channel Case* in 1949; although the UNCLOS equates this with the changes carried out in the Law of the Sea, such as those concerning the breadth of territorial waters or the appearance of new maritime areas such as the exclusive economic zone or the waters of archipelago.

However, the UNCLOS did not just retake the expression "straits used for international navigation", but it is also presented as the heir to the 'definition', or better the 'indefinition' of the expression throughout its article. In the words of J.N. Moore, "UNCLOS has not altered or clarified the existing uncertainty in customary international law over the definition of 'straits used for international navigation'".[133]

In fact, the Convention does not define what straits used for international navigation are and, therefore, does not specify which straits come within the scope of Part III; likewise it fails to establish the differentiation between straits which are used for international navigation and those which are not.

There is no doubt that this accumulation of circumstances means that it is necessary to carry out the previous, complex task of properly conceptualising this legal shape which is the subject of our study, insofar as this is possible. Although we must point out that the name, the *terminology*, is not the decisive factor in the definition of straits, as shown by the fact that areas which clearly come within the category of international straits are termed "channels" (*St. George's Channel, North Channel, Santa Lucia Channel,...*), or "passages" (*Martinique Passage, Aruba-Paraguana Passage, Jacques Cartier Passage,...*). Besides the particular denomination which they might receive in a determined language, as occurs with "belt" in Danish (*Litlle Belt, Great Belt*), "proliv" in Russian (*Proliv Karskiye, Proliv Friza,...*), "kaikyo" and "suido" in Japanese (*Tokara Kaikyo, Tsugaru Kaikyo, Shikotan Suido, Taraku Suido,...*) and "selat" in Malayan–Polynesian (*Selat Sunda, Selat Lombok, Selat Alas,...*). Thus, this aspect has no judicial importance.

The same occurs with the *maps* as regards whether a determined strait is identified or not. These do not have probative value in this respect; and constitute mere tools of an auxiliary nature.

3.1 Some Proposals of Definitions Put Forward Within the Framework of the III Conference

The representative of the German Government said in his reply to the request for information from the Preparatory Codification Committee of The Hague of 1930 that "the notion of straits has not yet been fixed in Public International Law".[134]

---

[133]Vid. Moore (1980, p. 112).

[134]Cited in Gidel (1934, p. 729).

This statement continues to be in force after the adoption of the Geneva Convention of 1958 which, although, as we have seen, it establishes a specific regime for the straits used for international navigation in its article 16.4, it does not contain definition of these.[135] And, although it may seem strange (especially if we take into account the repeated doctrinal vindication concerning the need to fill this gap), it still remains unresolved by the UNCLOS.

It is true that, throughout the III Conference of the United Nations on the Law of the Sea, the advisability of including a definition in this regard was discussed, although the difficulty involved in achieving a concept that would be generally accepted was also pointed out. In this regard, several explicit and implicit proposals of a definition of the "straits used for international navigation" were presented.

One proposal, which was co-sponsored by ten Arab States, referred to the straits which communicate two parts of the high seas and are habitually used for international navigation.[136]

The proposal submitted by Canada was more precise, as Canada was especially interested in expressly including the reference to habitual use for international navigation; its intention was to exclude the *Northwest Passage* from this classification, a group of straits only used sporadically and seasonally. In this respect, Canada defined an international strait in the following terms:

> An international strait is a natural passage between land formations which:
> (a) (i) Lies within the territorial sea of one or more States at any point in its length and
> (ii) Joins two parts of the high seas...
>
> (b) Has traditionally been used for international navigation.[137]

The British proposal stated that the articles on the passage of straits used for international navigation should be applied to any straits or other water course in the geographical sense, which are used for international navigation and connect two parts of the high seas; the proposal of the "Private Group of States" contained an almost identical description.[138]

[135]Despite the fact that the 1958 Convention does not include a definition related to the straits used for international navigation, the subject was addressed throughout the 1958 Geneva Conference. In this respect, Portugal proposed the addition of the words 'and sea routes' to the term 'straits', while Holland directly proposed substituting the term 'straits', by 'sea routes'; subsequently, a joint proposal of the United Kingdom, Portugal and Holland included the expression 'straits or other sea routes'. The reference to 'or other sea routes' was finally erased at the request of the prospers themselves. Cf. *United Nations Conference on the Law of the Sea, 1958, Official Records*, vol. 3, pp. 223, 225, and 231, respectively for each proposal.

[136]The proposal was presented by Saudi Arabia, Algeria, Bahrain, U.A.E., Iraq, Kuwait, Libya, Qatar, Syria and Tunisia. Cf. Doc. A/CONF.62/C2/L.44, August 7, 1974.

[137]Vid. Doc. A/CONF.62/C.2/L.83, August 26, 1974.

[138]Cf. Doc. A/CONF.62/C.2/L.83, August 26, 1974, for the proposal made by Great Britain. And Doc. A/CONF.62/C.2/L.3, July 3, 1974, for the Private Group of States.

However, neither the articles project co-sponsored by Malaysia, Morocco, Oman and Yemen, nor the text submitted by the "Group of coastal States of Straits", contained any reference to the geographical factor, and stressed the fact that the straits affected were the ones included in the territorial sea of one or several States, that is to say, the legal factor.[139]

Finally, the Conference decided not to include a definition in the text of the Convention, and, as we stated above, this endorses its Part III to the "Straits used for international navigation", but it did not specify what these terms meant at any point.

In this regard, L.M. Alexander stated that the expression "is so inexact as to render the phrase itself almost meaningless";[140] a circumstance which constitutes a clear controversial factor as there may be a number of interpretations and a lack of essential precision.[141]

In such a situation, we are clearly addressing a problem of interpretation – one of the many which the UNCLOS gives rise to, as we will see – which must be resolved through recourse to the interpretational criteria included in article 33 of the 1969 *Vienna Convention on the Law of Treaties*, as this is a treaty authenticated in several languages; its bona fide nature, the text, the context and the current meanings of the terms are the first instruments of interpretation, not forgetting the complementary recourse to the preparatory work and the circumstances in which it was drafted. It is a question of adopting the common sense interpretation, as "a strict (or 'mechanical') interpretation could well be found to be inappropriate in certain settings".[142]

In this respect, we understand that the content of the proposals submitted by the States throughout the III Conference, referred to above, constitute a fundamental aspect of this clarifying work. Bearing these proposals in mind, it is possible to state that, to a greater or lesser extent, these are all tributaries of the three factors which has *traditionally* constituted the doctrinal definition doctrinal of 'international straits' or 'straits used for international navigation': geographic, legal and functional.

As stated by Hart, "a better approach seems to be that of seeking a 'concept' which consists of criteria, rather than a 'definition'".[143]

This leads us to conclude that the precise conceptualisation of the "straits used for international navigation" must include the determination of the geographic, legal and functional factors so that, in order to be able to identify an international strait,[144] these three components must be employed as a type of Holy Trinity.

139Cf. Doc. A/CONF.62/C.2/L.16, July 22, 1974, and Platzöder (1983, p. 187), respectively.

140Vid. Alexander (1987, p. 480).

141Cf. Momtaz (1974, p. 843).

142Cf. Nandan and Anderson (1989, p. 166).

143Vid. Hart (1961, p. 3).

144Etymologically, the term strait comes from the Latin "strictus". For a terminological analysis of the word 'strait' see, Brüel (1947, pp. 15–17).

3.2 The Geographical Factor

According to E. Brüel, a strait in the geographical sense is understood to be "a contraction of the sea between two territories, being of a certain limited width and connecting two seas otherwise separated at least in that particular place by the territories in question".[145] Based on this concept and following this author, we can state that there are four features or defining geographical characteristics of an international strait: a natural sea passage, a contraction of the sea, separating two land areas, and uniting two areas of the sea.

(a) In the first place, it must be a *part of the sea not created artificially*. Thus, the adjective '*maritime*' distinguishes the straits from other water channels which do not form part of the sea, as is the case of rivers which constitute fluvial areas. While the adjective '*natural*' differentiates these from other maritime water courses artificially created by man, such as canals, for example the Kiel, Suez or Panama Canals, which are subject to the absolute sovereignty of the States they cross, unless there is an agreement otherwise. Evidently, rivers and artificial canals do not fall within the category of straits.[146]

As regards this final aspect, R. Lapidoth[147] puts forward an interesting question on what the author classifies 'semi-natural passages', that is to say, the case of strait which becomes navigable as a result of excavation work carried out by a State. Is this subject to the legal regime of straits? The ICJ made a pronouncement on this particular point with regard to the *Fisheries Case* (United Kingdom vs. Norway), in relation to the legal classification of *l'Indreleia* – a typical geographical formation of the 'skjærgaard', literally a coast full of farallones and rock walls –, a navigation route in the interior of the Norwegian straight, base lines, stating that:

> The Court is bound to observe that the Indreleia is not a strait at all, but rather a navigational route prepared as such by means of artificial aids to navigation provided by Norway. In these circumstances the Court is unable to accept the view that the Indreleia, for the purposes of the present case, has a status different from that the other waters included in the 'skjægaard'.[148]

R. Lapidoth[149] holds that, unlike the majority opinion which understands that the Court has given a negative response to the inclusion of these semi-natural passages within the regime of straits, this conclusion cannot be reached with this extract.

We should remember that the case in question involved passages located in interior waters of Norway. The Court was not called on to pronounce on the application of the regime of straits to these passages, but on the thesis of the United

---

[145]Vid. Brüel (1947, pp. 18–19).

[146]For an adequate distinction and conceptualisation of these three waterways, see the interesting work of Baxter (1964, especially pp. 3–21).

[147]Cf. Lapidoth (1972, p. 32).

[148]Vid. ICJ *Report*, 1951, p. 132.

[149]Cf. Lapidoth (1972, p. 32).

Kingdom that these passages are located in territorial waters and not in interior waters; on rejecting this thesis, the Court specified that it was going to analyse the question "for the purposes of the present case",[150] that is to say, from the categorisation of the waters as interior or not. This specification and the particular circumstances of the case enable us to conclude with R. Lapidoth that the decision of December 18, 1951 in the *Fisheries Case* has not laid down a general rule in relation to semi-natural passages.

From our point of view, we consider that the response to this interesting question is found in the practice of the States. In this regard, we observe that there are many important straits recognised as such – for example, the *Turkish* Straits, the Straits of *Malacca* and the Straits of *Singapore* –, were the subject of artificial, technical work in order to make them more navigable and it has never been suggested that they have ceased to be straits and become artificial canals. In addition, if each navigation route enabled due to the execution of coastal, technical work is withdrawn from the regime on straits, this will enable the 'Straits' States to exclude a strait from this regime as they wish.[151]

There is also the case of the appearance of a new strait, that is to say, a passage which arises or becomes navigable as a result of an earthquake. We consider that, in this case, the natural character of the strait as an essential requirement of the geographical component is fully complied with, which would entail its inclusion in this category on condition that the other components required are present.

In any case, we understand that the navigable or non-navigable character of a strait and, consequently, the execution of artificial work on the coast in order to make a natural passage navigable, or to make it more safely navigable, do not influence the geographical component which will always exist when there is a natural passage of maritime waters; this particular aspect will only occur in relation to the functional component which we will refer to below.

(b) In the second place, there must be a *contracting of the sea* in such a way that the water in the straits, compared with the adjacent waters, is more reduced in extension than these. This aspect is very imprecise as, from a geographical point of view, there is no specification of the width the contraction must have. The determination of the maximum width is important for the purposes of the inclusion of the strait as an international strait although this question is part of the legal component.

In addition, the geographical point of view is only limited to the superficial aspect of the strait and does not take other factors into account such as the depth.

(c) Furthermore, it must *separate two land areas*, and it is indifferent whether these are two continental territories (*the Strait of Gibraltar*, *Bab El Mandeb*, *the Strait of Hormuz,...*), two islands (*the Menorca Channel*, *the Strait of Bonifacio*, *the Cook Strait,...*), or a continental territory and an island (*the Strait of Messina*, *the Strait of Corfu*, *the Channel of Zanzibar,...*).

[150]Vid. ICJ *Report*, 1951, p. 132.

[151]Cf. Schwarzenberger (1957, pp. 340–341).

However, we believe that it is important to point out that, although it is irrelevant as regards the definition of straits, the type of territories separated do have ***legal consequences*** as concerns the navigation regime applicable to the straits, for example, in the case that mainland and island territory of the same State are separated with more than one option of passage, the navigation regime is different as the right of passage in transit will not be in force, but the right of innocent passage which cannot be suspended, as we will see below.

(d) Finally, the straits must *unite two sea areas*, two maritime areas where the straits are the natural connection between both. The size of the waters connected is not a determining factor either. These may be two oceans – *inter ocean straits* – (*the Magellan Strait, Le Maire Strait, the Torres Strait,...*), two parts of the same ocean – *intra-ocean straits* – (*the Bass Strait, the Robenson Channel, the Strait of Juan de Fuca,...*), two seas – *inter-sea straits* – (*the Sound, the Western Chosen Channel, the Straits of Makassar,...*), or two parts of the same sea – *intra-sea straits* – (*the Strait of Messina, the Kalmar Sound, the Strait of Otranto,...*). A fifth category of Straits are those which join an ocean with a sea – *ocean-sea straits* – (*the Strait of Bering, Bab el Mandeb, the Strait of Gibraltar,...*).

As regards this last note and as in the previous case, an important ***legal question*** arises concerning the union between the geographical and legal components and the proof of the interdependence between the defining components of an international strait. This involves the nature of the maritime areas connected by the waters of the straits. In this regard, we find that most of the doctrine analyses this matter exclusively from the geographical point of view. However, we do not share this point of view as we understand that the classification of the marine areas connected is a question, which, together with its geographic aspect, falls fully within the legal area as it is International Law and not geography which defines and delimits which spaces may be considered to be the high seas, exclusive economic zones, territorial seas or interior waters of the States.

As regards this matter, the ICJ in the *Corfu Channel Case* set out the decisive criteria for the definition of an international strait as its situation as a connection component between two parts of the high seas.[152] However, the 1958 Geneva Convention extended the areas connected referring to its article 16.4, not only to those which join a part of the high seas, but also those where the straits connect the high seas with the territorial seas of a foreign State; which gives rise to two passages. There are no doubts as regards the customary nature of the first type of passage, which connects two areas of the high seas, but it is not very clear whether this nature can be duly applied to the second type of passage. In fact, this second case does not appear in the preliminary work of the ILC and was incorporated at the end in order to promote the Israeli claims as concerns a navigation in the Gulf of *Aqaba* and through the Strait of *Tiran*.[153]

---

[152] Cf. ICJ *Reports*, 1949, p. 28.

[153] Cf. Dean (1958, p. 623).

In the opinion of R. Lapidoth,[154] there is no doubt that this involves a customary norm and its affirmation is founded on several declarations made by maritime powers. However, this does not take into account the many contrary declarations made by other States – especially the Arab States[155] –, stating the exclusively conventional nature of this rule; nor the fact that neither doctrine nor case law had referred to these types of passages before 1958.[156] Therefore, it is inadvisable to classify the norm included in the 1958 as customary as regards this second type of passage.

Evidently, when the III Conference and the adoption of the UNCLOS took place, this polemic was overcome. In this regard, the range of waters connected by international straits is again extended in the 1982 Convention, and logically includes the new maritime areas which arise in the III Conference of the United Nations on the Law of the Sea, such as the exclusive economic zone (EEZ) and archipelagic waters. Consequently, several types of passages can be distinguished. One is the classical strait which connects two areas of the high seas which is included in the news EEZ area and is a connection between the high seas and the EEZ and the high seas or the EEZ (article 37). The polemic kind concerning the connection between the high seas or the EEZ and the territorial waters of another State was also codified (article 45).

However, two more cases were added: the straits which separate an island from a mainland State, when the other side of the island has a similar high sea or EEZ corridor (article 38); and the exceptional case of the connection of the interior waters of a State, in the hypothesis that the tracing of the straight base line has resulted in enclosing waters which were not considered as such as interior waters (article 35(a)).

Although it is not expressly include in Part III, it may be inferred that the stipulations in Part IV regarding the "Archipelagic States", which in the case of international straits between the islands of an archipelagic State, that is to say, in the case of "Archipelagic Straits" the waters of these straits, as well as the maritime areas connected by these are basically archipelagic waters of the coastal State of the

[154]Cf. Lapidoth (1972, pp. 19–22).

[155]According to Ambassador Shukari, "the amended text no longer dealt with general principles of international law, but had been carefully tailored to promote the claims of one State" (vid. Doc. A/CONF.13/C.1/L.39, 1958). At the 14th Meeting of the Second Committee of the III Conference, the representative of Kuwait, speaking on behalf of Iraq, the United Arab Emirates, Libya, Saudi Arabia, Qatar and Kuwait, confirmed that the States on whose behalf he spoke were not part of the Geneva Convention on territorial sea and contiguous zone de 1958, as this provision "had been politically motivated by the desire to accommodate specific interests in a particular region" (vid. Doc. A/CONF.62/C.2/SR.14, 1974).

[156]Thus, when the International Law Association adopted the Rules of the Sea in 1895, it did not refer at any time to the straits which connected the high seas with the territorial sea of a State, nor are there any references at the 1930 Conference of The Hague, nor in the Court Decision of 1949. In fact, its inclusion was due to the proposal of a British amendment submitted during the Conference in 1958, as a solution to the controversy concerning the Strait of Tiran, as stated above (Momtaz 1974, p. 844).

Straits on condition that this State is an archipelagic State and has delimited its maritime areas with archipelagic baselines (article 47). This occurs especially with Indonesia (*Bangka Passage*, *Strait of Lombok*, *Selat Alas*, *Strait of Sunda*,...) and the Philippines (*Balintang Channel*, *San Juanico Strait*, *Strait of Maqueda*, *Strait of Polillo*,...); although we also find straits of this type in the Bahamas (*Crooked Passage*, *Northeast Providence Channel*, *Northwest Providence Channel*), Trinidad and Tobago (*Galleons Passage*), Papua New Guinea (*Vitiaz Strait*, *Dampier Strait*,...), Salomon Islands (*Indispensable Strait*, *Manning Strait*,...), and Fiji (*Vatu-i-ra Channel*). Although, in this last case of 'archipelagic straits' the legal regime applicable to these is not that of Part III, but that of Part IV.

In line with this last observation and in relation to the distinct nature of the maritime waters connected by the 'international straits', we must finally note the existence of a gap in the framework of Part III in particular, and in the UNCLOS itself in general. This lack refers to the possibility that a strait connects an area of the high seas or EEZ with the archipelagic waters of a State. This is not a hypothesis, it is a reality. This situation arises in the *Wetar Strait*, between Timor Leste and Indonesia, which connects the EEZ or high seas of the Sea of Arafura with the Indonesian archipelago waters.

3.3 The Legal Component

It is usually established that there are several hundred straits on the planet in the geographical sense of the term explained above, but only certain number of these are included in this category of the Law of the Sea. In order to be included, certain legal aspects which we have mentioned above and which are balanced between the geographical and the legal must be complied with, such as those related to the land surfaces separated, and the maritime spaces connected, and a specific legal component, namely, that the waters of the strait form part of the territorial sea of one or more States. It is not necessary that the totality of the waters of the straits are included in the territorial sea, it is sufficient that overlapping occur in some parts of the strait.[157]

This component is implicit in the Geneva Convention of 1958 as it is included in the part dedicated to the territorial sea, and more specifically in section III dedicated to the right of innocent passage of foreign vessels through the territorial sea. Years before, several authors, such as J. Westlake[158] had pointed this out at the beginning of the twentieth century, when he stated that the waters of the strait form part of the territorial sea of on or more States as the Straits considered in International Law were those whose width is less than double the extension of the territorial sea of the coastal States of the strait in question. A thesis which was included in the proposals

[157] Cf. Pharand (1977, p. 66).
[158] Cf. Westlake (1904, p. 193).

of the Preparatory Committee of the Hague Conference of 1930, and included in the Conference itself when the Principal Committee pointed out that the waters of the straits which do not form part of the high seas are territorial seas.[159]

For legal purposes, therefore, an international strait is a geographic strait in which the territorial sea of the coastal State or States does not permit the existence of a freely navigable through this seaway.[160] If this is not so, in the event that double the width of the territorial sea is exceeded, that is to say, that the width of the strait exceeds 24 nautical miles in all its parts (on condition that the width of the respective territorial sea is 12 miles, as, if it is less, the maximum limit referred to would be proportionally reduced), this would be geographic strait but not an international strait as part of its waters would not be submitted to the sovereignty of any coastal State and there would be a freely navigable route and freedom to fly over the strait as it would be high seas or an EEZ.

It is important to stress, therefore, that there is a direct relationship between the extension given to the territorial sea and the number of straits considered as such for legal purposes; only when the width of the strait takes in the sums of the territorial seas of the coastal States (regardless of the extension which have these have decided to attribute to themselves), the strait will be legal. If there is a band of high seas or an EEZ between these, it will not be legal.

In this regard, it will be necessary to take into account if there are **islands** in the strait belonging to one or more coastal States, which must inevitably be taken into consideration for the purposes of establishing the width of the straits and, therefore, its legal nature. The existence of islands in a strait may convert a purely geographic strait into a legal one. Evidently, these islands will have their own territorial seas attributed and these may overlap with the territorial sea of the mainland coast, which will mean that the waters of the strait are territorial sea and, therefore, the legal component which would not be complied with otherwise is verified.

This circumstance will obviously depend on the breadth of territorial sea which the coastal State decides to attribute to its mainland and to its islands located in the strait. If it chooses 12 miles, it will probably mean that the strait becomes a 'strait used for international navigation' in the sense of Part III. This occurred with the *Bass Strait,* between the island of Tasmania and the Australian mainland where there are about 50 islands; Australia's attribution of a territorial sea of 12 miles to all of these meant the transformation of this strait, traditionally 'free', into an 'international strait'. A circumstance which was repeated in the *Torres Strait*, between Australia an Papua New Guinea; the existence of islands belonging to both States, with their own territorial seas, reduced the minimum distance of the navigable channel in Australian territorial waters to 20 miles.

However, it may occur that the coastal State or States decide to attribute an area of territorial sea of less than 12 miles, which restricts this metamorphosis. This occurs, for example, with the *Kattegat*, strait of the North Sea which is over

---

[159]Cf. *AJIL*, 1930.

[160]Cf. Yturriaga Barberán (1993, p. 254).

24 miles wide and takes as reference points the respective mainland coasts of Denmark and Sweden; therefore, *a priori*, it does not comply with the legal component. However, there are two islands in this strait (the Danish islands of Laesø and Anholt) and if these are taken into consideration, they divide the strait into two channels; neither of the channels exceeds 24 miles in width. In fact, the width is reduced to 12 miles in the narrowest part between the Danish peninsula of Jutland and the island of Laesø, and is 22 miles wide in the east channel, between this island and Sweden. This circumstance would make the *Kattegat* comply with the legal component and, consequently, it would enter within the framework of application of Part III of the UNCLOS. However, faced with the common extension of the territorial seas to 12 miles, at the end of 1979, Denmark and Sweden agreed to attribute a territorial sea of 3 miles to each other, leaving a strip for free navigation in both channels of the strait amounting to 6 and 16 miles, respectively; which means that the provisions of Part III are not applicable to the *Kattegat*.

This also occurs in the Strait of *Kadet*, between Denmark and Germany, and in several Japanese straits.

In this regard, it should be pointed out that, in fact, the effect of islands on a strait is that it divides the strait into one or more channels which require the application of the legal component test in order to classify each of these as 'international straits'. Channels which rarely appear individualised as straits in practice. This individualisation, however, occurs in a certain number of cases. Such is the case of the Strait of *Korea* – between Korea and Japan –, where the Japanese island of Tsushima divides this into the *Western Chosen or West Korea Strait* (23 miles), and the *Tsushima* or *East Korea Strait* (25 miles); however, neither of these enter into the scope of application of Part III because there is a high sea route in these, as consequence of Korea's and Japan's decision to attribute themselves a territorial sea of 3 miles in this area.

Mention should also be made of the Strait of *Luzon* – between Taiwan and the Philippines –, divided into three channels as a result of the multiplicity of islands: *Bashi Channel* (53 miles), *Balintang Channel* (24 miles) and *Babuyan Channel* (14 miles); as a result of this individualisation, the last two mentioned are clearly 'international straits'.

This legal factor, which we refer to, is expressly mentioned in article 36 of the UNCLOS which stipulates the following:

> This Part does not apply to a strait used for international navigation if there exists through the strait a route through the high seas or through an exclusive economic zone of similar convenience with respect to navigational and hydrographical characteristics; in such routes, the other relevant Parts of this Convention, including the provisions regarding the freedoms of navigation and overflight, apply.

As examples of geographical and functional straits which clearly do not enter within the scope of application of Part III of the UNCLOS, as stated in article 36, because their widths exceed 24 nautical miles, we can cite: the *Strait of Otranto*, *Malta Channel*, *Skagerrak*, *Chetvertyy Kurilskiy Proliv*, *Strait of Florida*, *Yucatan*

Channel, *Caicos Passage*, *Anegada Passage*, *Strait of Taiwan*, *Balabac Passage*, and the *Mozambique Channel*, among many others.

Besides the traditional case we have referred to, which stipulates that the waters which form part of the straits are territorial seas, we must take into account another two *special situations* of international straits whose waters are not territorial seas, and which are also referred to in the UNCLOS.

Firstly, the particular case included in article 35(a) of the Convention concerning straits whose waters have become *interior waters* as a result of straight base lines whereas previous to these lines being drawn, they were not interior waters, that is to say, they were territorial waters or high seas; following the provisions in article 5.2 of the Geneva Convention of 1958. It is evident that, in this hypothesis, despite the applicability of Part III, the waters which form part of the straits continue to be interior waters and not territorial seas.

Secondly, there is the case referred to above of the 'archipelagic straits' included in Part IV of the UNCLOS. As we have seen, in this case the waters which form part of this special type of international straits are not territorial seas either, but archipelagic waters.

3.4 The Functional Component

Together with the geographical and legal components, the straits must comply with a final criterion of functional nature: the use of the straits for international navigation. A condition which, in turn, presupposes the intrinsic requisite that this seaway must be navigable, if it is not, evidently it would not enter the legal category being analysed at this point.

We must state that the navigation requirement refers exclusively to *maritime navigation* and not air navigation, which clearly does not have the navigability problems arising on the sea. In fact, in the light of an analysis of the doctrine on the subject, as well as the opinions of the States in this regard, except for the United States, for obvious reasons, the conclusion can clearly be drawn that, when reference is made to the use of a strait for international navigation or not, this is done thinking solely of sea navigation, and, *surface* navigation, not underwater. This is the most logical position if it is taken into account that we refer to a maritime area whose waters are territorial sea, where innocent passage referring exclusively to maritime surface navigation has traditionally been in force.

Until the arrival of the UNCLOS, the overflight of aircraft and underwater navigation were not allowed in territorial seas including straits. Thus the reference to the straits used for international navigation, codified in the conventional regime of 1958 and reproduced in the regime of 1982, must be understood with regard only to maritime surface navigation. In fact, the straits which are not used for maritime navigation or are not navigable do not give rise to overflight problems or, even that underwater navigation may be feasible. However, it is understood that they do not comply with the functional requirement, remain outside Part III, and are regulated by the general regime of Part II concerning territorial seas.

This is the case of the *Dampier Strait* in Papua New Guinea, located between the Island of Umboi and New Britain, which joins the Sea of Bismarck to the north with the Salomon Sea to the south, but is not navigable.

Three of the four straits which separate the Andaman Islands, a group of five islands which form part of the Indian territory of the Union of the Andaman and Nicobar Islands, located in the Bay of Bengal; specifically, the *Strait of Austin* between North and Middle Andaman, the *Strait of Homfray* between Middle Andaman, Baratang and the northern point of South Andaman, and the *Strait of Andaman* between Baratang and South Andaman are not navigable. Only the *Strait of Macpherson* between South Andaman and Rutland Island is used for international navigation.

The Russian straits located in the Arctic, *Petukhouskii Shar*, *Sharapov Shar*, *Iugorshii Shar* and *Red Army* are not suitable for surface sea navigation, due basically to the intense and continual fog.

Furthermore, *McClure Strait* in the Canadian Arctic archipelago is not navigable either due to the fact that it is blocked by ice throughout the year, although some submarines can cross it submerged.

On this point, as we have stated, all these straits which form part of the territorial seas of one or more States and which, despite their geographical situations, are not navigable or are not used for international navigation, remain outside Part III and they are, therefore, subject to the provisions of the Part II of the 1982 Convention on the right of innocent passage which may be suspended in.

Undoubtedly, when the UNCLOS refers to "straits used for international navigation" it is expressly including the functional component which characterises an international strait, and copies the same formula used by the 1958 Geneva Convention, which is initially found in the *Corfu Channel Case*. In this case, the ICJ stressed the second determining criteria, although not the most decisive one, for the definition of an international strait (the fundamental criteria was the geographical one): the fact that the strait is used for the purposes of international navigation,[161] – maritime surface navigation.

But what does 'used for international navigation' mean exactly?

Undoubtedly, the question of the 'use' of the strait does not have mere theoretical relevance but is crucial and very difficult to interpret. While, the geographical and legal components are easy to determine, it is extremely complex to specify the functional usefulness as this is, in the opinion of S.C. Truver,[162] a very vague criteria. In fact, this is the component which generates the greatest discrepancy between States as those who use the straits always hold that this seaway is used for international navigation, while the coastal States challenge its international nature and reject such use, It occurred in the 1970s with the *Strait of Malacca*, which

[161] Cf. ICJ *Reports*, 1949, p. 28.

[162] Cf. Truver (1980, p. 150).

Indonesia and Malaysia refused to classify as international strait, and with Canada in relation to the *Northwest Passage*, which Canada holds is part of Canadian interior waters and is not open to international navigation.

We can also state that one of the main reasons which put the successful termination of the III Conference at risk was the fact that the majority of the 'straits' States were blinded by the desire to exclude their own straits from such use, and , therefore, from this definition.[163]

As pointed out by R.R. Baxter, "it is impossible to answer in the abstract how many straits meet this requirement of being 'useful' for the international navigation";[164] and the opinions of the doctrine in this regard are varied and even disparate. Therefore, the proposals such as those of C. Jiménez Piernas are not off the mark. In his opinion, this minimum criterion of usefulness must be corrected through the introduction of a number of objective factors, such as the number of vessels and flags, the tonnage and the value of the cargoes,[165] together with the use of such straits by warships.

In this same regard, E. Brüel points out that, in order to determine the importance of a strait, subjective criteria are not sufficient, but impartial considerations are necessary such as, "the number of ships passing through the strait, their total tonnage, the aggregate value of their cargoes, the average size of the ship, and especially, whether they are distributed among a greater or smaller number of nations".[166] All these factors are guidelines and none of them can be considered to be decisive. The fundamental point for this author is that the interest in the use of the strait has a world nature.

Similarly, Ch. de Visscher pointed out that the data used by the Court in the *Corfu Channel Case*, are not conclusive. In his opinion, the number of vessels, the total tonnage, the value of the cargo and the number and variety of flags represented are the four useful factors for determining the importance of a strait.[167]

J.D. González Campos has a similar opinion and states that the idea of usefulness expressed by the ICJ is satisfied with a minimum of special usefulness of the sea passage, which not only constitutes an essential or necessary route for maritime navigation, but is possibly a useful or convenient route.[168]

However, O'Connell[169] holds that mere potential usefulness is insufficient. An opinion shared by T. Treves.[170] Contrary to this, R. Lapidoth understands that a

---

[163]Cf. Platzöder (1978, p. 724).

[164]Vid. Baxter (1964, p. 9).

[165]Vid. Jiménez Piernas (1982, p. 796).

[166]Vid. Brüel (1947, pp. 42–43).

[167]Cf. Visscher (1969, p. 142).

[168]Cf. González Campos (1975, p. 306).

[169]Cf. O'Connell (1984, p. 314).

[170]Cf. Treves (1985, p. 789).

weak flow of traffic may be sufficient on condition that the strait can serve for navigation.[171]

Two very clear *conclusions* can be obtained from the opinions given here. The first is that there is no agreement in the doctrine regarding what must be understood by the expression "used for international navigation". The second is that there is no unanimity as regards the degree of use of a strait required for it to be considered as ***international***. This use has never been "quantified".

In our opinion, we understand that 'use' cannot be confused with 'importance'. Moreover, criteria such as those explained above: tonnage of vessels, value of the cargo, types of vessels, variety of flags, etc. do not refer to establishing the usefulness of a strait for navigation, but its importance its strategic, commercial, defensive or communications importance has no relevance whatsoever as regards the applicable legal regime. In addition, these criteria vary with time; therefore, they do not contribute legal security in this respect. The functional component does not distinguish between straits; it only requires that the strait be used for navigation, regardless of the type of use. The fact that the States prefer to use alternative communication routes does not subtract from the internationality of a strait which is navigated at some time.

It is evident that we are faced with a **problem of interpretation**. Thus, in order to reach a conclusion it is essential to bear in mind the rules of article 33 of the Vienna Convention on Law of Treaties of 1969 – which, in the end, refers to those set out in articles 31 and 32 –, and in particular the authentic texts of the UNCLOS because all of them attest.[172] Thus the French text refers to "Détroits servant à la navigation internationale", while the Spanish is similar to the English and uses the expression "Estrechos utilizados para la navegación internacional". Taking the current meaning of the terms "used", "servant" and "utilizados" as the first criteria of interpretation in accordance with the stipulations in article 31.1 of the Convention of 1969, in our opinion, it is clear that the verbal tense which the legislator wished to use is the 'present continuous', and in consequence that the use is not potential but real.

This conclusion is strengthened by recourse to another objective component of interpretation which is the very text of the treaty as component of the context (article 31.2 Convention of 1969). In this regard, within the Section dedicated to passage in transit, article 37 of the English text of the UNCLOS stipulates that, "this section applies to straits **which are** used for international navigation".[173] The use of the present tense is more than evident and, consequently, it is also evident that there is necessary that the use of the strait would be current, which corroborates our previous thesis.

[171]Cf. Lapidoth (1972, p. 31).

[172]The different authentic texts of the 1982 UNCLOS can be consulted at http://www.un.org/Depts/los/convention_agreements/convention_overview_convention.htm.

[173]Vid. *Ibidem*. The bold lettering is ours.

As regards the terms "which are" that, as we have seen, only appear in the English text, S.N. Nandan and Sh. Rosenne[174] hold that these introduce a *temporary* aspect to the functional criteria, in the sense that only those straits which are being used for international navigation at the time the Convention comes into force will be governed by passage in transit. We do not share this conclusion. On the contrary, we understand that the meaning and effect which must be granted to these terms is purely descriptive and 'atemporal'; which is also confirmed by the versions in other authentic languages.[175] The legal regime of Part III will be applied to the international straits which, at any time, comply with the three components we have stated above. The disappearance of any of these at any time would exclude a strait from the category analysed here, in the same way that if a purely geographical strait in the future complies with the legal and functional components, it would fall within the category of international strait and, therefore, Part III of the UNCLOS would apply to it.

In addition, we consider that in this clarifying work must also retain the preparatory work of the Convention, as a *complementary interpretation resource* which confirms our previous conclusion (article 32 of the 1969 Convention). In this regard, several proposals were presented at the III Conference in order to include the term "customarily used".[176] As we saw above, Canada proposed a definition of a strait which included that it "has traditionally been used".[177] Chile was also in favour of the 'traditional use'.[178] However, none of these proposals was finally accepted. In contrast, when referring to the right of passage through archipelagic waterways, the article 53.4 of the UNCLOS uses expressions such as "normal passage routes", or "normal navigational channels". The inclusion of these references in Part IV and the omission of 'normally' in Part III[179] means that it is not

[174]Vid. Nandan and Rosenne (1993, p. 290).

[175]In fact, the Preparatory Committee wanted to delete the terms "which are" from the English version, as well as "que sean" from the Spanish version; however, only the Spanish words were changed to "utilizados (used)", in line with the other versions, while the English text remained unchanged (cf. the reference in George 2002, p. 194).

[176]This is the case of Algeria and others (Doc. A/CONF.62/C.2/L.44), and of Iraq (Doc. A/CONF.62/C.2/L.71).

[177]Cf. Doc. A/CONF.62/C.2/L.83.

[178]This was done in the Second Committee, on July 23, 1974 (cf. *Official Records*, vol. 2, p. 138, 14th meeting).

[179]We believe that, in relation to the Geneva Convention of 1958, it is equally important to recall that, at the 1955 and 1956 meetings, several members of the ILC proposed alternative expressions such as, "straits vital for international navigation" or "straits essential for international navigation" (cf. ILC, *Yearbook*, 1956, vol. I, p. 203). The final project of the Commission presented at the First Conference of the United Nations on the Law of the Sea in 1958 decided on the expression "straits normally used for international navigation". As we pointed out in the first chapter, the term 'normally' was, however, finally suppressed from the text of the Convention in the course of the 1958 Geneva Conference as a result of an amendment cosponsored by Holland, Portugal and the United Kingdom, which argued that the ICJ had not used the expression 'normally' to refer to navigation (cf. Doc. A/CONF.13/C.1/L.39, of March 25, 1958).

necessary that this be a normal channel, but that *any type of navigation use can be relevant in order to classify a strait as international.*

In fact, no evidence of use may be excluded *a priori* as the *exceptional* or non-traditional use. Both new and remote use can be taken into consideration. The use may be *civil or military, or both.* The use does not have to be regular nor reach a predetermined level,[180] in such a way that, subsidiary straits which are not essential for international navigation may be included in Part III. It is true that, when the authors refer to 'international navigation', they are normally referring merchant navigation, but we must also consider military navigation. In fact, the important changes introduced by the UNCLOS into the regime of international straits regarding the transit of the passage mainly benefit surface warships, military aircraft and submarines.[181]

Thus, we agree more with the opinion of H. Caminos, who states that "as long as some degree of use actually exists, any strait connecting two parts of the high seas or an exclusive economic zone with another, or the high sea or exclusive economic zone with the territorial sea of a foreign State, can be classified as 'a strait used for international navigation'".[182]

Evidently, as we pointed out, at any time it may occur that a strait included in Part III *ceases to be used*, or *cannot be used* (because it becomes non-navigable) for international navigation. In such a case, this maritime area will not comply with the functional component, thus it would cease to belong to the category of 'straits used for international navigation', and, consequently, the provisions of Part III would not apply to it, however those of Part II regarding the territorial sea and the right of innocent passage with the possibility of suspension would.

Moreover, the supposition may be put forward inversely. It is possible that a strait is not used for international navigation or is not navigable, but it does have the geographical and legal components, *and becomes navigable and starts being used for international navigation.* On this point, one inevitably must think of the technology applied to navigation. The technological progress can lead to better, more powerful ships with more sophisticated technology which enables them to navigate through straits where navigation was impossible before, and the functional requirement is not complied with. Logically, if this occurs, this strait would become an 'international strait' as from this point and would fall within the scope of application of Part III of the UNCLOS (a position which, as we stated above, is counter to what is maintained by H.B. Robertson, as we have seen above).

---

[180]Cf. Nandan and Anderson (1989, p. 169).

[181]In fact, with the passage in transit, surface warships, military aircraft and submarines will have freedom to navigate and fly over the strait for the purposes of rapid uninterrupted passage. While, in the regime of innocent passage, as we will see in Chap. 5, many States demand previous notification or authorisation for surface warships to access their territorial waters, including the straits; aircraft have no right to fly over the territorial seas and submarine navigation is prohibited.

[182]Vid. Caminos (1972, pp. 128–129).

This hypothesis has arisen recently as regards the aforementioned *Northwest Passage*,[183] a group of straits which constitute the east-west route between the Atlantic Ocean and the Pacific Ocean through the Canadian Arctic Archipelago, and especially since 2007, because that year the ice around the North Pole has reduced to the minimum and a maritime route lying the Pacific and the Atlantic Ocean has been open. Since its waters are frozen over throughout almost the whole year, this passage is not usually used for international merchant navy navigation, except for a short period of time (2 or 3 months a year in summer) when it is navigable although normally with the aid of icebreakers. In this regard, as pointed out by D. Pharand, there are five routes in the *Northwest Passage*.[184] Only two of these are really feasible for the navigation of deep draught vessels in arctic class 7 or more, which require a minimum of 20 m depth; and what we will call Route 1 is better than Route 2 because the climatic conditions are less severe.

Specifically, Route 1 runs along the following seaways: *Lancaster Sound* (between Bylot and Devon Islands, with a width of 51 miles), *Barrow Strait* (between Prince Leopold and Devon Islands, with a width of 15.5 miles), *Viscount Melville Sound*, (between Victoria and Melville Islands, with a width of 66 miles), *Prince of Wales Strait* (between Banks and Victoria Islands, with a width of 6.5 miles), and the *Gulf of Amundsen* (separates the mainland from Banks and Victoria Islands, with a minimum width of 56 miles). As we have pointed out, this is the most feasible route for the icebreakers and was chosen by the *Dome Petroleum Company* as the main route. Route 2 is an alternative to Route 1, it follows the same direction as far as *Viscount Melville Sound*, but then continues through the *McClure Strait*, which separates Banks Island from Prince Patrick and Melville Islands, and has a minimum width of 43 miles; this Strait is wider that the *Prince of Wales Strait* in Route 1, but, as we pointed out, it is blocked by ice throughout the year, which makes it impossible to navigate.[185]

---

[183]We should point out that this same approach can be applied to the so called *Northeast Passage*, on the Arctic coast of Russia, which has many points in common with the *Northwest Passage*, fundamentally focused on the dialectic between inclusion in or exclusion from the category of international straits in relation to the straits which exist within these passages; a dilemma which is polarised around the navigable nature or not of these straits and the legal nature of the waters involved.

[184]Vid. Pharand and Legault (1984, pp. 1–21); a work to which we refer for a detailed study of this route and each of the straits which compose it.

[185]The other three routes noted by Pharand and Legault (1984, pp. 1–21), have shallow waters, therefore, they are only suitable for small boats with shallow draughts. These routes are the following:

ROUTE 3: *Lancaster Sound*, *Barrow Strait*, *Peel Sound* (between the Prince of Wales and Somerset Islands, its width is 15 miles), the *Franklin Strait* (between Prince of Wales Island and the Boothia Peninsula on the mainland, and has a width of about 20 miles), and *Larsen Sound* (the continuation of the previous one). From this point, there are two options: One, continue through the *Victoria Strait* (between Victoria and King William Island, with a minimum width of 50 miles), the Gulf of *Queen Maud*, the *Dease Strait* (between Victoria Island and the Peninsula of Kent, with a width of 12 miles), *Coronation Gulf*, and the *Straits of Dolphin and Union* (15 miles wide); the other involves continuing through the *James Ross Strait* (a passage

The proof of the operability of Route 1 as regards the technical plan was provided in 1969, when the American oil tanker 'Manhattan', weighing 151,000 tons and horse power of 43,000, crossed the Northwest Passage for the first time, accompanied by an ice breaker.[186] More recently, the 'MSFram' started the September 24, 2009 a cruise from Oslo to New York, crossing Greenland and the fiords of Canada.

This has led to serious discrepancies as regards the classification of the *Northwest Passage* in general, and the series of straits which make it up in particular, as 'international strait'.[187] Today the majority doctrine excludes the whole of the *Northwest Passage* from the classification as an international strait because it cannot be used for international navigation. A doctrine which is preached and fostered by the coastal State of these straits, Canada, due to evident strategic and defensive interests, which are vital interests for the States which are omnipresent in the classification of international straits.

However, this situation could change in the not too distant future, and make this debate a sterile one if the current global heating of the planet and the resulting melting of the arctic zones, currently the layer of ice is 35% thinner than three decades ago. In this sense, in summer of 2007 the Northwest was fully open to navigation and the ice layer has reduced to 4.24 millions of square kilometres, The *Intergovernmental Panel of Climate Change* of United Nations upholds that in the 2040s the Arctic would be without ice in the summers.[188] If this happens, the *Northwest Passage* would become a navigable maritime channel, and would undoubtedly be much used for international navigation taking into account the

---

30 miles wide, between King William Island and the Boothia Peninsula), *St. Roch Basin*, the *Rae Strait* (between the Gibson Peninsula on the mainland and the King William Island, 13 miles wide), *Rasmussen Basin*, and the *Simpson Strait* (separates the Adelaide Peninsula from the King William Island, with a minimum width of 2 miles).

ROUTE 4: This only varies as regards route 3 in that, instead of crossing *Peel Sound*, from the *Barrow Strait* it continues through the *Prince Regent Inlet* (between Somerset Island and Brodeur Peninsula on Baffin Island, with a width of 40 miles) and the *Bellot Strait* (between Somerset Island and the Boothia Peninsula, with a width of little over 1 mile) in order to enter the *Franklin Strait*.

ROUTE 5: This is another alternative to route 3, which varied at the beginning as, instead of beginning at *Lancaster Sound*, it begins in the *Hudson Strait* (between the Ungava Peninsula and Baffin Island, with a width of 60 miles), *Foxe Channel* (between Southampton Island and the Foxe Peninsula on Baffin Island, 80 miles wide), *Foxe Basin*, the *Fury and Hecla Straits* (between the Melville Peninsula and Baffin Island, its narrowest part is 0.5 mile) and the *Gulf of Boothia*; from this point it joins the *Strait of Bellot*.

[186] Cf. Rigaldies (1987, p. 49). A description about the evolution of the passage through the *Northwest Passage* to date can be consulted in the work of Krafft (2009, pp. 537–578).

We must point out that this has not been the only incursion made by United States vessels into the *Northwest Passage* in order to demonstrate the feasibility of its navigation, such as the nature of its maritime routes as international straits. Thus, in 1985, the icebreaker *Polar Sea* crossed these waters without the permission of Canada; as did United States nuclear submarines in 2005.

[187] As regards this particular point, see, Pharand and Legault (1984, pp. 110–121).

[188] Cf. http://www.ipcc.ch.

fact that it considerably reduces the distance and, therefore, the time required to reach the Atlantic from the Pacific: A voyage from Tokyo to London involves 15,000 miles by Cape Horn, 12,600 miles by the Panama Canal, 8,600 miles by the Suez Canal (which only admits small ships), and 7,900 miles by the *Northwest Passage*.[189]

There would be no doubts about compliance with the functional component which is the subject of discussion today. The response to the question put forward by D. Pharand, in the sense that, if international navigation takes place "would this result in the internationalisation of the passage?" and, more specifically, "will the number of foreign flags and foreign transits be sufficient to put the Northwest Passage in the category of straits being used for international navigation?"[190] Our response is categorical: Yes, and we believe that time will prove us right.

As concerns this last case that a non-navigable strait may become navigable and be used for international navigation, a final consideration may be put forward as regards the fact that this circumstance occurs thanks to the *navigation aids* or the technical work carried out by the coastal State of the strait. This question would require differentiating the type of aid. Thus, these aids which do not have to be facilitated by the coastal State, but may be provided by the State which uses the strait while it navigates it; this is the case of the icebreakers which cross the arctic straits. Evidently, in this case, the fact that the aid given in order to navigate the strait does not have to be provided by the coastal State means that the functional component is not affected, nor is the international nature of the strait called into question.

The situation arises when it is a question of technical aid or work which can only be carried out by the coastal State. As occurs with the two lighthouses placed by Australia in the eastern area of the *Bass Strait* (on Deal Island and on King Island) in order to make navigation feasible on this route, which is quite complicated and difficult. The coastal State does not have any obligation to carry out this aid, if it does not do so the strait will not be navigable, it will not comply with its functional component and, consequently, it will not fall within the category of 'international strait' nor will the provisions of Part III be applicable to it. However, if the coastal State facilitates this aid or work so that the strait is navigable and it begins to be used for international navigation, from our point of view, from this time the coastal State of the strait is obliged to maintain this aid or work due to the application of the legal institution of 'estoppel'; that is to say, it has been 'prevented', it is bound by its own acts, although the coastal State refuses to see it as such. We can say that the functional component is complied with and that the strait in question passes to the category of 'strait used for international navigation'.

---

[189]Cf. Macnab (2004, pp. 1–2).

[190]Vid. Pharand and Legault (1984, p. 110).

3.5 Finalising a Concept of 'International Strait'

From the above, it is possible to conclude a final concept of what must be understood by 'strait used for international navigation', which is to say, an 'international strait'. This would be, *"a natural maritime passage which entails a contraction of the waters no greater than double the width of the territorial sea of the respective coastal States, which separates two land masses, and communicates a high seas or EEZ area with another high seas or EEZ area, or a high seas or EEZ area with the territorial sea of another State or, possibly, with its interior waters or its archipelago waters, and is used for international navigation"*.

The subsequent question is inevitable. Considering the hundreds of straits in the geographic sense which exist on the planet and which in fact appears in geophysical studies,[191] which of them are subject to the rules of Part III of the UNCLOS? Or, in other words, **which are the international straits in the world**?

The response is far from easy and is extremely complex, especially for a jurist as other factors of an exclusively geographic and geo-morphological nature are added to the purely legal factors, such as the corresponding widths, the pertinent coordinates, or the navigation channels, which are beyond the mind of the lay person as regards this matter. Besides these points there are differences and contradictions between papers on the crucial question of the minimum width in each of the straits. For example, this occurs with the *Guadeloupe Passage*, which, according to the Kennedy Report,[192] is 20 miles wide, while according to Lay, Churchill and Nordquist,[193] the width is 28 miles. The same happens with the *Strait of Kara*, to which Lay, Churchill and Nordquist[194] attribute a width of 19 miles, while it appears in *The Times Atlas*[195] with a width of 29 miles, and W.E. Butler[196] reduces this to 13.5 miles at its narrowest channel. Such a variation may make a determined strait oscillate between the balance of the legal and the merely geographical depending on the source of reference.

Another equally conflictive aspect refers to the effect the existence of islands may have on the establishment of the width of a strait, a point we have already referred to. When determining the minimum width of a specific strait, at times neither geographers nor jurists take into account the existence of islands in the strait belonging to one or both coastal States, and only take the respective mainland coasts of the States of the strait as points of reference. However, as we pointed out,

[191]In *The Times Atlas and Encyclopedia of the Sea* (London, 2nd ed., 1983, pp. 150–151), the figure is raised to 274. In this regard, the list of straits contained in the virtual Encyclopaedia: http://en.wikipedia.org/wiki/List_of_straits may be useful and is easy to access and use.

[192]Cf. Kennedy, R.H., "Estudio geográfico e hidrográfico de los estrechos que son rutas de navegación internacional", Doc. A/CONF. 13/6 (23 October 1957), p. 3.

[193]Cf. Lay, Churchill and Nordquist (1973, p. 886).

[194]Cf. *Ibidem*, p. 890.

[195]Cf. *The Times Atlas and Encyclopedia of the Sea* (1983, p. 150).

[196]Cf. Butler (1978, p. 19).

on occasions there are islands throughout the State, islands which have their own territorial sea, therefore they must be taken into consideration and, consequently, this would considerably reduce the minimum distance between both sides of the strait, especially if the sovereign coastal State decides to attribute 12 mile territorial waters to these.

Among other examples, this situation arises in the *Strait of Florida*, between the United States and Cuba in the Caribbean Sea. However, in this case, although the width of 82 miles commonly facilitated[197] does not take the aforementioned islands into consideration, if these were taken into account, this would not entail a change as regards the legal regime, bearing in mind that that the location of these continues to give a width greater than 24 miles, that is to say, a strait not included in Part III.

However, in other cases, the existence of islands in the strait with its own territorial sea reduces the width of the strait to such an extent that straits initially not included in Part III due to exceeding the width of 24 miles would become international straits.

Among others, this is the case of the *Barrow Strait* in the Canadian artic archipelago, which has a width of 27 miles according to studies made.[198] However, along the length of the strait there are several islands (Griffith, Garrett, Lowther, Young, Hamilton), which divide the strait into several channels, none of which exceed 16 miles in width – in fact, the minimum width is 8 miles and the maximum 15.5 miles. Thus, if we take these islands into consideration and grant them a territorial sea of 12 miles, the legal effect is that all the waters of the *Strait of Barrow* would become territorial seas of Canada and, consequently, the strait would be included in Part III of the UNCLOS.[199]

The same occurs with the aforementioned *Bass Strait*, in Australia. The width usually attributed to this strait is approximately 100 miles,[200] in consequence it would be out from Part III. However, this figure obviously does not take into account the chain of islands which are abundant in the eastern area of the strait, from Tasmania to Victoria, which has its own territorial sea, and converts the waters of the eastern entry of the strait into a territorial sea and, therefore, into an 'international' strait. It should also be mentioned that this is the case of the *Torres Strait*, which we have also referred to. Or the *Bering Strait*, between Russia and the United States, which is usually excluded from the list of international straits as it is

---

[197]Cf. *The Times Atlas and Encyclopedia of the Sea* (1983, p. 150). On December 16, 1977, the U.S.A. and Cuba signed an agreement delimiting the maritime frontier, which took into consideration all the islands we refer to, in order to delimit their respective territorial seas and their exclusive economic zones, and it is clearly shown how the territorial seas of both States do not overlap despite the existence of islands, with their own territorial seas (cf. the text of the agreement in *Limits in the Sea*, No. 110).

[198]This appears in *The Times Atlas and Encyclopedia of the Sea* (1983, p. 150).

[199]We must not forget the polemic arising with regard to the Canadian artic straits concerning their navigable nature or not, referred to above.

[200]In *The Times Atlas and Encyclopedia of the Sea* (1983, p. 151) it is attributed a minimum width of 80 miles; while *Wikipedia* raises this figure to 119 miles.

considered to exceed 24 miles in width, without taking into account the two islands in the centre of the channel, Big Diomede which belongs to Russia, and Little Diomede which is under U.S. sovereignty, which are separated by scarcely 2 miles of sea, and these divide the straits into two channels, whose widths are no greater than 19 miles; thus, the eastern channel is a territorial sea of the U.S.A., while the western channel is a territorial sea of Russia.

All these obstacles make it difficult to achieve an exhaustive catalogue of international straits, bearing in mind the non-existence of an 'official' inventory of such straits.

It is true that a large number of legal works on the straits consulted refer to a map drawn by the Office of the Geographer of the Department of State of the United States of America,[201] which gives a figure of 116 straits which would be affected by the extension of territorial waters to 12 miles. However, while we do not wish to question the precision and importance of this work, we must bear in mind the words of the ICJ in relation to the value of cartography "as evidence of an auxiliary or confirmatory kind", and "they cannot be given the character of a rebuttable or 'iuris tantum' presumption".[202] Following the pronouncement made by the Court, only those maps incorporated into treaties, which are the expression of the will of the States will have legal force. Moreover, the map of the United States Department of State does not take into account other data which as we have seen can exclude the inclusion of a strait in Part III, such as the category of waters which connect or the functional requirement; or the possible existence of an EEZ or high seas corridor, when the coastal States do not extend their territorial waters to 12 miles. Therefore, the value of this map is merely referential.

Furthermore, **the number varies considerably** in other works consulted. Thus, L.M. Alexander[203] gives a figure of 265 straits used internationally for navigation. Smith[204] considers 220 straits to be international. While the list provided by Lay, Churchill and Nordquist[205] places the figure at 136. Larson[206] considers that there are 134 international straits. The French National Commission on Toponimy cites less than a hundred in its "Nomenclature of Maritime Areas".[207] However, Brüel[208] mentions 35 important straits. While, the report prepared for the First Conference

[201]Cf. US Department of State, Office of the Geographer, *Map of World Straits Affected by a 12 Mile Territorial Sea*, No. 564375, Dec. 1974 (this is a repetition of the one drawn up in 1971, No. 510376).

[202]Vid. *Frontier Dispute Case* (Burkina Faso / Mali), ICJ *Reports*, 1986, pp. 583, 56.

[203]Cf. Alexander (1987, p. 480).

[204]The list is in Koh (1982, pp. 24–26).

[205]Cf. Lay, Churchill and Nordquist (1973, pp. 885–891).

[206]Cf. Larson (1979, p. 56).

[207]Cf. Commission Nationale de Toponymie, *Nomenclature des espaces maritimes*, de 2-11-2004, pp. 5–13; this is a document drawn up within the framework of the project of the International Hydrographical Organisation, entitled "Limits of Oceans and Seas".

[208]It also mentions another five straits of secondary importance which would also be international straits, which gives a total of 40. Cf. Brüel (1947, pp. 44–45).

on the Law of the Sea by the frigate captain R.H. Kennedy[209] describes 33 straits used as navigation routes and whose width does not exceed 26 nautical miles; a number which is increased to 34 by R. Lapidoth.[210]

In the light of this oscillation of figures, we fully share the opinion of O'Connell in the sense that:

> The statistics are a little **misleading**, because many of these straits are dubiously 'international' in the sense of being habitually used by shipping as required by the principles laid down by the International Court of Justice in the Corfu Channel case, and indeed find themselves in the list only because of assiduous cartographical scrutiny on the part of those who seek to produce **arresting figures**.[211]

Therefore, it is not surprising, as Beckett points out, "that the works of jurists do not contain comprehensive lists of international straits, since they are concerned to expound principles and not to provide catalogues".[212]

This does not mean that we don't wish to accept the challenge involved in identifying the 'international straits', but simply point out the limits which this task entails and, therefore, the fact that it is impossible to have a complete identification. Thus, the inclusion of the 'straits used for international navigation' in each of the categories included in Part III of the Convention,[213] which we will deal with in the following chapter, is not intended to be an exhaustive list.

[209]Cf. Kennedy, R.H., "A Geographical and Hydrographical Study of the Straits which are International Navigation routes", Doc. A/CONF.13/6 (October 23, 1957).

[210]Cf. Lapidoth (1972, pp. 121–122) (she adds the Strait of Tiran).

[211]Vid. O'Connell (1975, p. 97) (the bold lettering is ours).

[212]Vid. *Corfu Channel Case*, ICJ *Pleadings, Oral Arguments, Documents*, 1949, pp. 549–550.

[213]A list of international straits, located by geographic areas, with the reference of the relevant article in Part III (or Part IV in some cases) which establishes the category it belongs to, can be consulted at the end of this book.

Chapter 4
The Standardised Catalogue of International Straits Presented in Part III of the 1982 United Nations Convention on the Law of the Sea

Together with the 'imprecision' which we have seen arises from the concept of straits used for international navigation, and which adds confusion to a regime which is already quite confusing, the 1982 Convention was responsible for generating other examples of imprecision when it broke up this regime based on a surprising multiplicity of 'types' of international straits, each one with a different legal regulation. One "diversification" which is the fruit of one of the major commitments of the III Conference, which was graphically expressed with the axiom: "No strait, no fish".

Coherent with the legal nature of the waters of the strait and as we have seen in the first chapter, the 1958 Geneva Convention consolidated the unity of the legal regime in all the territorial seas including the straits. A unity of regime which mades no distinctions between straits used for international navigation. However, this unity and homogeneity was broken with the 1982 Convention which not only established autonomous regulation for the straits, but diversified this in a range of different legal regimes, each of which was applicable depending on the category of strait involved within the varied catalogue of the UNCLOS.

Along the same lines, article 16.4 of the 1958 Convention expressly distinguished two types of straits: those which unite two parts of the high seas, and those which unite an area of the high seas with the territorial sea of a State, both governed by the right of innocent passage. A third category should be added to these and is implicitly deduced from the provisions of article 5.2, and the right of innocent passage was also applicable to this: international straits which, as a consequence of the establishment of a straight baseline, now include areas which once were territorial seas or the high seas as internal waters. However, the Geneva Convention does not refer to the straits with conventional 'long-standing' legal regimes (such as the Turkish Straits, Danish Straits and the Magellan Strait), which, as we saw in the first chapter, were historically the first to have a precise, agreed, legal regime. Evidently, the omission did not prevent the particular conventional regime stipulated from governing the straits.

A.G. López Martín, *International Straits*,
DOI 10.1007/978-3-642-12906-3_4,

However, in the UNCLOS, a type of 'multiplication' of the categories of straits took place, and each strait was linked to a different legal regime, and a 'metamorphosis' came about.

The situation was as follows.

(A) There are a number of STRAITS WHICH ARE BEYOND THE SCOPE OF APPLICATION OF PART III OF THE UNCLOS. Namely:

1. *Straits which include internal waters which, before the establishment of a straight baseline in accordance with the method established in article 7, were already internal waters* (article 35 a).
2. *Straits whose passage is regulated in whole or in part by long-standing international conventions in force specifically relating to such straits* (article 35 c).
3. *Straits whose passage is regulated by agreements compatible with the Convention which are not 'long-standing'* (article 311.2).

   Straits through which there is a high sea route or which are crossed by an exclusive economic zone, also advisable as regards hydrographical and navigation characteristics (article 36).
4. *Straits which include archipelagic waters as a consequence of the drawing of archipelagic lines in archipelagic States* (part IV).

(B) There are another two types of straits to which the RIGHT OF INNOCENT PASSAGE is applied:

1. *A strait formed by an island of a coastal State of this strait sand its mainland when there is a high seas route on the other side of the island or it crosses an exclusive economic zone, also advisable as regards hydrographical and navigation characteristics* (article 38.1).
2. *Straits situated between one part of the high seas or of an exclusive economic zone and the territorial sea of another State* (article 45.1 b).

(C) Finally, a third category of straits, the most numerous and the one which includes the most important straits, from the economic, strategic and military points of view, where the RIGHT OF PASSAGE IN TRANSIT governs:

1. *Straits between a part of the high seas or of an exclusive economic zone and another part of the high seas or of an exclusive economic zone* (article 37).

(D) We believe that it is necessary to point out that, despite the plurality of straits typified, not all are included. In fact, in spite of the variety, no fewer than eight categories of straits, practice has shown that there are two types of straits used for international navigation which are not included in any of the categories mentioned above; these are *OMISSIONS* of the Convention which give rise to serious doubts and discrepancies concerning the legal regime which must be applied to these.

1. Firstly, there are those which could be called *MIXED STRAITS*, that is to say, straits which belong to two categories of straits with different legal regimes, which are mutually incompatible. Specifically, *straits formed by an island of a coastal State, on one side, and its mainland and the mainland of another coastal State, on*

the other side, when there is a high seas route on the other side of the island or it crosses an exclusive economic zone, also advisable as regards hydrographical and navigation characteristics. That is to say, the straits in article 38.1 but with another coastal State in the straits, in such a way that, if only the part of the strait which affects the latter is taken into account, this would involve a strait which joins a high sea zone or an exclusive economic zone with another high sea zone or exclusive economic zone.

The *Corfu* Strait (between the Greek island of Corfu and the Greek and Albanian mainland) and the *Pemba* Channel (between the Tanzanian Island of Pemba and the mainland of Tanzania and Kenya),[214] are examples of straits involving this dichotomy; which we have termed 'mixed straits'.

Does article 38.1 apply and, therefore, the right of innocent passage applies, or, as there is another coastal State is this provision not pertinent and article 37 applies, that is to say, the right of passage in transit? The solution is unclear and doctrine is divided one way or another, with those who defend one or other solution.[215]

From our point of view, we understand that it is more logical (at least harmful for the sovereign coastal State on both sides of the strait) to apply the right of innocent passage, therefore, the application of article 38.1, especially if we bear in mind that there is an alternative free passage on the other side of the island. In this regard, it is important to retain the origin of the category of article 38, that is to say, the proposal of Italy, and the discussions of this at the III Conference. Although it was designed with the Straits of *Messina* in mind, according to J.A. Yturriaga Barberán,[216] its authors hoped to satisfy other States such as Tanzania, the coastal State of the *Pemba* Channel, a mixed strait. If this was the intention of the authors, should it be taken into consideration as a component of interpretation?

Furthermore, the strict literal meaning of article 38.1 states the existence of a strait formed by an island of "a" coastal State of the strait; that is to say, it uses the indefinite article "a", and not the definite article "the", and this indefiniteness does not exclude the possibility that there is "another" coastal State which does not have sovereignty over the island. This exclusion would have occurred if the expression used by the UNCLOS had been "the coastal State", or the expression "exclusively formed" would have been used in the drafting. However, this is not so. We also

---

[214]Through an exchange of notes in 1976, Tanzania and Kenya agreed to the delimitation of their respective territorial seas, both as regards their lateral border and their border facing each other on the Island of *Pemba*. The text of the treaty, as well as an illustrative map clearly showing the border between Kenya and the island of Pemba (Tanzania), can be consulted at the Office of the Geographer, US Dept' State, *Limits in the Seas*, No. 92.

[215]Thus, among others, Treves (1985, p. 789–790) defends its inclusion in the case of article 38.1; while Caminos (1987, p. 139–140) holds the contrary opinion.

[216]See Yturriaga Barberán (1991, p. 292).

understand that article 38.1 really wishes to exclude the cases in which the island does not belong to any of the mainland States of the straits.

In any case, there are no doubts regarding the drafting of this provision, which, in our opinion, allows for the inclusive interpretation in article 38.1 of those straits composed of an island of a coastal State of this strait and its mainland territory when there is another coastal State on the mainland.

Another curious piece of information must also be taken into account. The dialectic only arises as regards the Corfu Strait, while the inclusion of the *Pemba* Channel in article 38.1 is unanimous, even for authors such as H. Caminos, who discard this in the case of *Corfu*. What is the justification of this discriminatory treatment?

In any case, it is clear that the Convention does not have a specific response to this; it involves a new question concerning interpretation.

2. In addition, the case of the *straits which include territorial seas and connect a part of the HIGH SEAS or an exclusive economic zone with the ARCHIPELAGIC WATERS of another State*. This situation occurred after East Timor became independent, in the Wetar Strait between Indonesia and East Timor,[217] which joins the high seas or the exclusive economic zone of the Arafura Sea with Indonesian archipelagic waters. What is its legal regime? The UNCLOS does not stipulate this. Faced with this 'gap', we understand that the legal regime applicable should be the right of innocent passage, by analogy with the category included in article 45 b): straits located between a part of the high seas or an exclusive economic zone and territorial seas.

4.1 Straits Excluded from the Scope of Application of Part III of the UNCLOS

When establishing the scope of application of Part III, article 35 operates only exclusively and establishes two categories of straits which the provisions of this part do not apply to: the straits formed by internal waters and those regulated by 'long-standing' conventions. Article 36 also excludes the straits used for international navigation crossed by high sea route or an exclusive economic zone. In addition, taking all the UNCLOS into consideration, we have located another two categories of straits which remain outside Part III: the archipelagic straits and those whose passage is regulated by an international treaty compatible with the Convention.

[217]Previous to independence, achieved after the referendum held on August 31, 1999, although total independence was not achieved until May 20, 2002, East Timor had been a province of Indonesia since 1975, as a result of its illegal military occupation. Thus, this strait was totally included in Indonesian waters and therefore was an 'archipelagic' strait.

Thus, there are five categories of international straits which are outside the scope of application of Part III[218] but, at least three of these are not outside the Convention as their regulation would be subject to other parts of the UNCLOS. This is the case of straits formed by internal waters which would be governed by the stipulations in Part II, on the straits through which there is a high seas route or an exclusive economic zone to which Parts II, V or VII would apply depending on the zone, and of the archipelagic straits regulated by Part IV. While the other two types of straits will be regulated by specific international treaties, a clear application of the principle of *lex specialis*.

4.1.1 Straits Which Include Internal Waters Which Previously Were Neither Territorial Seas Nor High Seas

The article 35 UNCLOS establishes:

> "Nothing in this Part affects:
>
> a) any areas of internal waters within a strait, except where the establishment of a straight baseline in accordance with the method set forth in article 7 has the effect of enclosing as internal waters areas which had not previously been considered as such".

It is evident that this provision refers to a particular type of strait, such as the straits in internal waters, and the traditional doctrine had excluded this type of strait from the general regime on straits,[219] but UNCLOS had decided to make a statement on these. If we pay careful attention to the provisions in this article, we will find that the Convention took into account two legal situations regarding these types of straits, depending on the legal base of the status of these waters. That is to say, there are two *different classes of straits in internal waters*, each with a different legal regime although, in both cases, the legal nature of these waters is not affected.

(a) There are *straits which are formed by internal waters which have always been internal waters*. This category of straits remains outside the scope of application of

---

[218]It is clear that these straits which are territorial seas but are not navigable, or are not used for international navigation do not come under the application of Part III as they are not part of the general category of "straits used for international navigation", that is to say, they are not 'international straits', their legal regime would be that of Part II. This aspect was fully analysed in the previous chapter when addressing the functional component of the concept of "straits used for international navigation", therefore, we refer the reader to this explanation.

[219]Lapidoth (1972, p. 23) rejected completely the application of the regime of straits which connect internal waters, arguing that not only the nature of the waters but the fact that they are not used for international navigation; while both these arguments seem to us to be futile today. Guiliano (1975, p. 16–17) had a similar opinion; or Solodovnikoff (1980, p. 331) also. This last author was undoubtedly inspired by the fierce Russian position which considered all its straits to be internal waters in order to exclude international navigation.

Part III according to article 35 a). The effect which generates this provision is that the rights of transit or innocent passage of the vessels or aircraft in those parts of the straits which were territorial seas does not extend to the internal waters of the straits. These zones are regulated by Part II of the Convention, consequently, they are subject to the exclusive sovereignty of the coastal State.

As regards these types of straits, there is no right of passage and they cannot be considered to be 'international straits'.

When do such circumstances arise? When can we speak of internal waters which have not arisen as a consequence of the establishment of a straight baseline in accordance with the method of article 7? This possibility that a strait might include internal waters which have always been of this type, that is to say, they have not been transformed into internal waters as a consequence of the establishment of a straight baseline, may occur, as pointed out by D. Pharand,[220] as a consequence of the existence of *historic entitlements*. This would involve the hypothesis of historic waters which would create a type of 'historic straits' similar to the 'historic bays' referred to in article 10.6 of the Convention.

Practice provides some examples of historic bays which are fully recognized, such as Chesapeake Bay, Delaware Bay and the Gulf of Fonseca.[221] However, the situation differs as regards the existence of historic internal waters in straits. Except for *Indreleia* in Norway, there is no generalised recognition of any other strait which includes historic waters.[222]

It is true that there are two States which have been claiming the classification of historic waters for the waters of some of their straits. This is the case of Canada and Russia as concerns the *Northwest Passage* and the *Northeast Passage*, respectively. Both States sustain that the straits included in these routes are historic internal waters, therefore, they are not international straits, there is no right of passage, and they are subject to the exclusive legislation of the coastal State.

In consonance with this vindication, Canada dictated an Order on the Geographical Coordinates of the Territorial Sea on September 10, 1986,[223] according to which all the waters of the Canadian Arctic Archipelago are internal waters of

[220]See Pharand and Legault (1984, p. 98).

[221]We must point out that that, besides these cases of recognised historic waters, not straits, by the coastal States, but these are not recognised by third party States, therefore, it cannot be definitively stated that they are of such a nature. Among others, this is the case of the Estuary of the River Plate by Argentina and Uruguay, of the Bay of Ungwana by Kenya, of the Gulf of Sidra by Libya, and of the Gulf of Taranto by Italy, claims protested by the United States (cf. *Limits in the Sea*, No. 112, pages 10–12).

[222]There is also the possibility that *Long Island Sound* involves historic waters although, rather than a strait, this sea route must be considered to be a bay. Thus, at least, North American Case Law has declared this in the cases of *Mahler v. Transportation Co.*, of 1866, and *U.S. v. Maine* in 1983, as well as the report of the Special Relater of the Supreme Court in 1983 (see the reference in Jia 1998, p. 68–69).

[223]Cf. The Law in http://www.un.org/Depts/los/LEGISLATIONANDTREATIES/regionlist.htm.

Canada as they consider them to be internal waters. Similarly, Russia dictated likewise in its Decree of July 20, 1957 and in the Report of July 21, 1964.[224]

Likewise, India and Sri Lanka have claimed Palk Bay to be a historic bay, and the waters of the Palk Strait as historic waters. This claim was agreed to in the Treaty of June 28, 1974, and included in Law No. 41 of June 1, 1979 by India as by Sri Lanka in a Proclamation of January 15, 1977.[225]

It is well known that there are two requirements for classification as historic waters: the effective, continual exercise of sovereignty by the coastal state or the coastal States (the Gulf of Fonseca involves three States: El Salvador, Honduras and Nicaragua) of the waters in question and the recognition of this situation by other States.

In the three cases mentioned, there is no doubt that the first requirement stated is complied with. However, there is doubt about compliance with the second requirement, the recognition of this situation by the other States as the three vindications have been protested. Thus, the member States of the European Union made a joint declaration, in a note on July 9, 1986,[226] in which they rejected the existence of a historic entitlement which justified the Canadian delimitation, which they understood to be contrary to International Law. The United States has repeatedly protested the three claims as lacking grounds and being contrary to International Law.[227]

Undoubtedly the most peculiar claim concerning historic waters was the one made on December 24, 2003, by Ukraine and the Russian Federation regarding the Sea of Azov and the Kerch Strait. In fact, in a joint Declaration of the President of Ukraine and the President of the Russian Federation, both States have claimed that "historically the Sea of Azov and the Kerch Strait are internal waters of Ukraine and Russia" and, consequently, "military vessels flying the flags of other States can only enter the Sea of Azov and cross the Kerch Strait by invitation of Ukraine or of Russia agreed to with the State".[228] In this case, it does not seem that there are any of the parameters required for the proclamation of historic waters, therefore, such a declaration, subjecting the passage of foreign military vessels to the consent of the coastal States, is a clear infringement of the right of passage in transit which should be in force in the Kerch Strait in accordance with the provisions in article 37 of the Convention of 1982, which both States are parties to.[229]

---

[224]Cf. http://www.un.org/Depts/los/LEGISLATIONANDTREATIES/regionlist.htm.

[225]Cf. http://www.un.org/Depts/los/LEGISLATIONANDTREATIES/regionlist.htm.

[226]See the reference and a detailed analysis of the claims of Canada and Russia, in Scovazzi (2001, p. 69–84).

[227]Cf. "United States Responses to Excessive National Maritime Claims", *Limits in the Sea*, No. 112.

[228]See *Law of the Sea Bulletin*, No. 54, 2004, p. 117.

[229]Ukraine filed its ratification instrument on July 26, 1999, and the Russian Federation on March 12, 1997.

(b) However, if the *waters of the strait are internal as a consequence of the establishment of straight baselines*, in such a way that these were not considered as such, then this strait would be classified as the second type of straits formed by internal waters which do come under the scope of application of Part III. This means that, although the legal nature of these internal waters remains intact, the straits continue to be 'international' and, depending on the case in question, the right of innocent passage or of passage in transit will apply in accordance with the provisions of Part III.

On this point, we disagree with M. George,[230] who stated that article 35 a), explicitly excludes the application of passage in transit in these types of straits. In our opinion, what this legislation really does is refer these types of straits to the provisions of Part III, which means that it will be necessary to examine the types of sea zones they connect in order to relocate them in one of the categories which expressly come under the scope of application of this part and, consequently, determine whether the application of innocent passage or of passage in transit are legal.

As regards these types of straits, article 35 a) carries out an assimilation of the internal waters to the territorial only for the purposes of passage.[231]

As we stated above, this same solution was implicit, to a certain extent, in article 5.2 of the 1958 Geneva Convention, when it established the right of innocent passage for those zones which had become internal waters as a result of the straight baseline. It is evident that this provision was also applicable to the straits because the Geneva Convention, unlike the UNCLOS, did not grant them an autonomous regime, but included them in the general regime of the territorial sea and this is a provision which forms part of this general regime.

A provision which is identical to this article 5.2 can be found in article 8.2 of the UNCLOS.[232] Thus, what is the reason for article 35 a)? According to B.B. Jia,[233] this article was necessary because, otherwise, the international straits would have lost their 'status'. In fact, if article 35 a) did not exist, these types of straits which include internal waters as a result of the establishment of straight baselines, whereas previously they had not been so, would come under the scope of application of article 8.2 of the UNCLOS, that is to say, in the general regime of territorial seas of Part II and, consequently, all of these, indistinctly, would have the right of innocent passage.

Similarly, within the framework of the 1958 Geneva Convention, there was no need for an express article similar to article 35 a), taking into account that the straits came under the general regime of territorial sea and the right of innocent passage

---

[230]Cf. George (2002, p. 193).

[231]Cf. Pharand and Legault (1984, p. 91).

[232]Article 8.2 of the UNCLOS provides that, "Where the establishment of a straight baseline in accordance with the method set forth in article 7 has the effect of enclosing as internal waters areas which had not previously been considered as such, a right of innocent passage as provided in this Convention shall exist in those waters".

[233]Vid. Jia (1998, p. 8–9).

was applied, thus, the provisions of article 5.2. were sufficient. Nevertheless, this situation changes in the UNCLOS where the straits are no longer regulated by the general regime on territorial seas, but they have their own autonomous regime. The legal nature of the internal waters, *a priori*, excludes the straits which enclose them from the concept of "straits used for international navigation" and, therefore, from Part III. Thus, an express provision was required to re-conduct these types of straits to the category of 'international straits' and to Part III of the UNCLOS in order to apply passage in transit or innocent passage to them depending on the case. As was well pointed out by O'Connell, "it was inconceivable that the major naval Powers would tolerate the potential closing of the more important straits upon the ground that they are internal waters".[234]

Finally, the expression used in article 8.2 and in article 35 a): "in accordance with the method set forth in article 7", must be understood in the sense that both provisions are applicable to the straight baselines established in the past and to those which will be established in the future on condition that the method established in article 7 is followed.[235]

Since 1951, a number of international straits have been closed as a consequence of the establishment of straight baselines, and such closures increased as from the seventies, with a tendency which is directly proportional to the extension of the territorial sea to 12 miles.[236] Whereas the right of innocent passage with the possibility of suspension was applied to the first straits, in accordance with the article 5.2 of the 1958 Geneva Convention, currently all of these have been re-conducted to the provisions in article 35 a) of the UNCLOS.

As we stated above, the list of straits which enclose internal waters continues to grow. Some examples of these are the following:

- **The United Kingdom** has included the following within its internal waters after the establishment of straight baselines through a Council Order of September 25, 1964:[237]
  - *The Pentland Firth*, between the Scottish mainland and the Orkneys.
  - *Little Minch*, between the Isle of Skye and the Outer Hebrides.
  - *North Minch*, between the Scottish mainland and the Outer Hebrides.

---

[234]See O'Connell (1982, p. 317).

[235]Cf. Nandan and Anderson (1989, p. 174).

[236]We will not go into the question of the validity of the establishment of these straight baselines which have included a number of international straits in internal waters although some of these have been protested by other States as they understood that they were not adapted to the delimitation criteria marked out in the UNCLOS. For example, this is the case of the delimitation made by Canada on its west, east and north coasts, protested among others by the United States in 1969, 1967 and 1986 (cf. *Limits in the Sea*, No. 112, p. 21). For a detailed analysis of the legal criteria which the States must adapt to for a delimitation of the straight baselines, see Office for Ocean Affairs and the Law of the Sea 1989.

[237]Cf. http://www.un.org/Depts/los/LEGISLATIONANDTREATIES/regionlist.htm.

- *The Solent*, between the Isle of Wight and the British mainland.
- In **Italy** the *Strait of Elba*, between the Island of Elba (which is part of the archipelago of Tuscany) and the continent, and the *Strait of Messina*, between Sicily and the continent, after the Presidential Decree of April 26, 1977.[238]
- In **Greece** among others are the *Corinth Channel* which communicates the Gulf of Corinth with the Gulf of Saronico and joins the mainland with the Peloponnesus; and the *Strait of Euripus* which separates the island of Euroboæa in the Aegean Sea from Boeotia on the mainland, in accordance with Presidential Decree 6/18, of June 3, 1931.[239]
- In **Croatia** the following straits included internal waters after the adoption of its maritime Code in 1994:[240]
  - *Bracki Kanal*, between Otok Brac and the mainland.
  - *Hvarski Kanal*, between Otok Brac and Otok Hvar.
  - *Korculanski Kanal*, between Otok Hvar and Otok Korcula.
  - *Lastovski Kanal*, between Otok Korcula and Otok Lastovo.
  - *Mljetski Kanal*, between Otok Mljet and the peninsula of Peljesac.
- **Montenegro** has internal waters in the *Strait of Nikola*, between Sveti Nikola and the mainland at Budva.[241]
- **Tanzania** has included the *Zanzibar Channel*, between the Island of Zanzibar and the mainland in its internal waters after the establishment of its Law on the Territorial Sea and the Exclusive Economic Zone of October 15, 1989.[242]
- The system of straight baselines established by **Oman** by Decree No. 38/82, of June 1, 1982, has included waters of the *Strait of Hormuz* as internal.[243]
- In **Japan**, after Cabinet Order No. 383 of 1993 and No. 206 of 1996 (June 20, 1996)[244] the following straits are in this situation:
  - *Rishiri-suido* between the Island of Rishiri and the peninsula of Wakana in Hokkaido.
  - *Sado-Kaikyo*, between the Island of Ryotsu and Honshu.
- **Taiwan** has included the *Pescadores Channel*, between the Island of Pescadores and Taiwan in its internal waters, in its Law of February 10, 1999.[245]

[238]Cf. *Ibidem*.

[239]Cf. *Ibidem*.

[240]Cf. *Ibidem*.

[241]Cf. *Ibidem*.

[242]Cf. *Ibidem*.

[243]Cf. *Ibidem*.

[244]Cf. *Ibidem*.

[245]Cf. *Limits in the Seas*, No. 127.

- In accordance with the Law of February 25, 1992 and the Order of May 15, 1996, **China** has included the following in its internal waters:[246]
  - *Lamma Channel* and *Tathong Channel*, between Hong-Kong and the Lamma Islands.
  - *Strait of Hainan or Quiongzhou*, between the Island of Hainan and the mainland.
  - *Bo-Hai Haixia* or *Strait of Po-Hai*, at the entry to the Bay of Pohai.
- By establishing straight baselines, in accordance with the Law of July 26, 1996, **New Zealand** has included the following as internal waters:
  - *Rangitoto Channel* and *Tamaki Strait*, small channels which communicate the adjacent islands with the city of Auckland inside the Gulf of Hauvaki.
  - *Cradock Channel*, *Colville Channel* and *Jellicoe Channel*, the three channels which connect the Gulf of Haurati with the Pacific Ocean.
  - *Floveaux or Foveaux Strait* between South Island and Stewart Island.
  - *Cook Strait* between North Island and South Island.
- **Canada** has a large number of straits included in internal waters besides those mentioned before which form part of the *Northwest Passage*, according to the establishment of the straight baselines stipulated by the Council Orders of October 26, 1967 (Labrador and Newfoundland), of May 29, 1969 (Nova Scotia, Vancouver and Queen Charlotte Islands), of May 9, 1972 (revision), and January 1, 1986 (Arctic islands).[247]
  - The *Chatham Strait*, the *Hecate Strait*, the *Dixon Entrance*, *Queen Charlotte Strait*, the *Georgia Strait*, the *Johnston Strait* and the *Brougton Strait*, in the Pacific.
  - The *Belle Isle Strait*, the *Jacques Cartier Passage*, the *Honguedo Strait*, the *Cabot Strait*, the *Northumberland Strait* and the *Canso Strait*, in the Atlantic.
- In **Chile**, after its Decree No. 416,[248] this situation involves the *Beagle Channel* which separates Tierra del Fuego from the Island of Navarino.
- Finally, we should refer to some of the straits which the **United States** (Proclamation No. 7219)[249] has in the Pacific and which include internal waters, such as the *Santa Barbara Channel*, between the Channel Islands and the mainland in California; or the *Golden Gate*, which connects the Bay of San Francisco with the Pacific Ocean. While in the Atlantic there is *Arthur Kill* or *Staten Island Sound*, which separates Staten Island from the mainland at New Jersey.

Do all these straits mentioned fall within the application of Part III? Evidently the answer is negative as their inclusion depends on their entering one of the cases referred to in this Part. If we take into account the parameters concerning 'coastal States' and 'connected waters', none of the hypotheses of Part III are included, and

[246]Cf. http://www.un.org/Depts/los/LEGISLATIONANDTREATIES/regionlist.htm.

[247]Cf. *Ibidem*.

[248]Cf. *Ibidem*.

[249]Cf. *Ibidem*.

these straits would remain within the scope of Part II; specifically, article 8.2 of the UNCLOS would apply, that is to say, innocent passage with the possibility of suspension. This happens when the straits are bordered by a single State and they join its own internal waters. In these cases, as Jia points out,[250] and following the practice of the United Kingdom and the United States in this regard, there would be an analogy with bays. These straits would be assimilated to *bays*. In this regard, we could cite the following:

- In **Russia**: *Orlovskaia* and *Gorlo* in the White Sea; *Vostochnaia, Zapadnaia, Anzarskaia* and *Zhizhginskaia*, in the Gulf of the White Sea.
- In **Croatia**: *Bracki Kanal* (between Otok Brac and the mainland), *Hvarski Kanal* (between Otok Brac and Otok Hvar), *Korculanski Kanal* (between Otok hvar and Otok Korcula), and *Lastovski Kanal* (between Otok Korcula and Otok Lastovo).
- In **Japan**: *Kii Channel*, between the Island of Shikoku and Honshu.
- In **China**: *Bo-Hai Haixia* or the *Strait of Po-Hai*, at the entrance to the Bay of Pohai.
- In **New Zealand**: *Rangitoto Channel* and *Tamak Straiti*, inside the Hauvaki Gulf.
- In **Canada**: the *Chatham Strait*, in the Pacific.
- In the **United States**: *Puget Sound*, the continuation of the *Strait of Juan de Fuca* towards the interior; the *Golden Gate*, in the Bay of San Francisco; *Arthur Kill* or *Staten Island Sound* and *Kill van Kull* which separate Staten Island from the mainland at New Jersey; *The Narrows* in Brooklyn; and *Rockaway Inlet* in Jamaica Bay.
- In **Chile**: the *Beagle Channel*.[251]

4.1.2 Straits Regulated Wholly or in Part by Long-Standing International Conventions

As provided in article 35 c), the second category of straits which is not regulated by the Convention refers to "straits in which passage is regulated in whole or in

[250] Cf. Jia (1998, p. 16).

[251] The inclusion of the Beagle Channel in Chilean internal waters derives from the arbitral decision of February 18, 1977, dictated by the Queen of England in the *Beagle Channel Case*, whereby the sovereignty of Picton, Lennox and Nueva Islands was attributed to Chile (the text of the decision can be found in *ILM*, vol. XVII, 1978, pages 632 et seq.). However, the decision was not accepted by Argentina until the conclusion of the Peace Treaty of November 29, 1984, signed thanks to the mediation of Pope John Paul II.

Article 13 and Annex No. 2 of the 1984 Treaty regulate navigation in the Beagle Channel specifically for Argentinean vessels, and generally for foreign vessels. In general terms, the regime stipulated is identical to the right of innocent passage with the possibility of suspension and subjection to the pertinent Chilean regulation (the text can be seen in http://www.un.org/Depts/los/LEGISLATIONANDTREATIES/regionlist.htm).

part by long-standing international conventions in force specifically relating to such straits".

In this case, there are a reduced number of straits which have their own special regimes which are safeguarded. It can be seen that this provision involves a recognition of acquired rights and a clear application of the principle of *lex specialis*.

However, this article poses some interpretational questions. One of these is the exact meaning of the expression 'long-standing'. What does the Convention precisely refer to with this expression? We consider that this is an unqualified expression which implies that, at the time the Convention was held, there were treaties which regulated the passage through some straits. However, the meaning of this provision, particularly, in combination with the terms "is regulated" and "in force", not only entails the idea that there was a conventional regime when the Convention was signed in 1982, but it also implies that the that the regulation of passage through these straits must have been in force for a long period of time although the UNCLOS does not specify how much time is required.[252]

On the other hand, the types of conventions which must regulate the straits are not specified, therefore, we understand that these may be bilateral or multilateral, and may be any type of treaty.

Furthermore, as can be deduced from article 35 c), there is a distinction between two types of regimes in force: those in which passage is totally regulated, and those which regulate the straits only partially. That is to say, these may be treaties which regulate all the aspects of passage or only regulate some aspects of passage through the straits in question. As regards this second category, a number of interesting problems arise; taking into account the fact that the long-standing convention in question regulates passage only partially, which legal regime regulates the questions related to passage which are not regulated in this convention? There are two possible interpretations.

One interpretation could be called 'preventive',[253] and stresses the proposition in article 35 "Nothing in this Part affects"; so that, in no case, will the legal regulation be the one contained in Part III. This means that those aspects of passage not regulated by long-standing Convention must be regulated by customary rules of International Law in force. Another possible interpretation is to understand the exclusion from the regime contained in Part III through article 35 c) refers only to those aspects regulated in the long-standing Conventions. This means that Part III would be applicable to the aspects not regulated by this convention.

While the first interpretation appears to be better adjusted to the literal meaning of the terms used, the second is the one which better responds to the intention behind the drafting of the autonomous regime of straits, that is to say, Part III of the Convention.

---

[252]Cf. Nandan and Rosenne (1993, p. 307).

[253]Cf. Schachte and Bernhardt (1993, p. 544).

Although the UNCLOS does not state the States this provision applies to, in the course of the Conference several delegations mentioned its applicability to the Turkish Straits, the Magellan Strait, the *Åaland Strait,* and the Danish Straits. The first three would be cases involving total regulation, while passage in the Baltic Straits is only regulated in part. Together with these cases generally recognised as examples of article 35 c), some authors have also included the Strait of *Gibraltar* in this context. They argue that the Franco-British Declaration in 1904, which prohibited the construction of fortifications along the Moroccan coast of the strait and guaranteed the freedom of navigation through this strait, together with the Treaty of November 27, 1912, constitute long-standing conventions.[254] The contrary opinion is the majority one and we share it. This opinion rejects the inclusion of the Strait of *Gibraltar* in the context of article 35 c). In this regard, the reference to the freedom of transit contained in these texts does not entail freedom of navigation, as in the case of the high seas, but the central component of the notion of innocent passage in the case of straits; that is to say, the non-prohibition of passage by the coastal State for reasons of security.[255] The 1902 and 1912 texts have, at least, a merely historic value while their opposition to third texts is null.

Therefore, only the four groups of straits mentioned above are excluded from the scope of application of Part III as they are totally or partially regulated by long-standing conventions. In all these cases, the regime on passage stipulated by these specific conventions is much more liberal than the one which would have corresponded by applying the provisions in Part III as it can be seen that, generally speaking, all these long-standing conventions establish a regime of freedom of transit through the straits they refer to.[256]

We first referred to the *Turkish Straits*. Passage through these was regulated by the Montreaux Convention of July 20, 1936.[257] These straits are:

- The *Bosphorus*, which is 28 km long (15 nautical miles) and, at its narrowest point, it is 600 m wide, and its maximum width is 3,700 m (2 nautical miles). It divides Istanbul into two parts. It joins the Black Sea to the Sea of Marmora.
- The *Dardanelles* is the continuation of the Bosphorus. It is 65 km (35 nautical miles) long, with a minimum width of 2 km (a little over one nautical mile), and a maximum width of 6,500 m (3.5 miles), it joins the Sea of Marmora with the Aegean Sea. Its depth is approximately 50 m.

The Strait of *Magellan*, with a length of 600 km (322 miles), is located on the southern border of Chile and Argentina; between the mainland and the Island of

[254]This minority opinion is defended by O'Connell (1984, p. 321) and Colombos (1967, p. 220).

[255]Similarly we can cite Caminos (1987, p. 131–132), Truver (1980, p. 179), Brüel (1947, p. 156), Baxter (1964, p. 335–336), among many others.

[256]The scientific literature which makes a detailed analysis of the legal regime of each of these long-standing conventions is very abundant; outstanding among these, and as an example, we draw attention to the work of Hill (1933, p. 477–556), Lapidoth (1972, p. 59–113), and Jia (1998, p. 109–125).

[257]See the text of the Convention in *AJIL*, 1937, sup. 1.

Tierra del Fuego and Cape Horn. It joins the Atlantic and the Pacific Oceans. The eastern approach lies between Cape Virgins and Espiritu Santo Cape, and its width varies between 22 and 12 miles. The western approach is situated between Desolation Island and the Evangelistas Islets (a group of rocks lying about 10 miles from the coast).

Its passage is regulated by article 5 of the Treaty on Limits made by Argentina and Chile on July 23, 1881, which stipulates that the Strait of Magellan will be open to the vessels of all nations.[258]

The *Åaland Strait* joins the Åaland Islands and the Sea of Åaland, in the vicinity of the Gulf of Bothnia, and is approximately 17 miles wide. This strait is regulated by the Swedish and Finnish legislation, and by a multilateral Treaty signed on October 20, 1921 on the demilitarisation and neutralisation of the Åaland Islands, which is based on the Treaty of Paris of 1856.[259]

The *Danish* or *Baltic* Straits have this denomination as they constitute three passes, which join the Baltic Sea and the North Sea through the *Kattegat* and the *Skagerrak*. Passage through these straits is partially regulated by the multilateral Treaty for the Redemption of the Sound, signed in Copenhagen on March 14, 1857, and by the bilateral Treaty of Washington signed on April 11, 1957.[260]

- The *Little Belt*, located between Jutland-Als and Fyn-Aerø, has a width which varies between 700 m and 27.5 km (15 miles), and a length of 50 km; its maximum depth is 75 m. Its islands divide the strait into several channels, and the one most used for navigation is the *Aarø Sund*, between the coast of Jutland and Aarø. The northern approach to the *Little Belt* is through the internal waters between Fyn and the islands of Samsøe and Endelave.
- The width of the *Great Belt*, between Fyen-Langeland and Sjaelland-Lolland, varies between 18.5 km (10 miles) and 28.2 (15.5 miles). At Sprogø it divides into two channels, both of which have similar widths (3.3 km).
- The *Sound* or *Sund,* (*Øresund* for the Swedes and *Sundet* for the Danes), is situated between Denmark (Sjaelland) and Sweden (Skåne). Its minimum width is 4 km (2.1 miles) and the maximum width is 47 km (25 miles). It is 58 miles long. At the eastern end, the deepest part is 26 feet. The Island of Ven divides it into two channels. One is the *Kattegat* with the Baltic Sea running north-south.

[258] Cf. *ILM*, 1984, p. 11 et seq. Both on signing and on ratifying the Convention, Argentina and Chile stated the application of the 1881 Treaty to the Strait of Magellan.

[259] Cf. *AJIL*, 1923, Supp. 1. In addition to this treaty there is the Moscow Agreement between Finland and the USSR signed on October 11, 1940. On signing the UNCLOS, both Sweden and Finland stated that the straits between the Finnish Islands and Sweden are included in article 35 c). However, this is the only case of the four which was the subject of a protest made by the United States, which is not a party to the 1921 Treaty and does not recognise that the Åaland Strait come under the scope of article 35 c) (cf. *Limits in the Sea*, No. 112, p. 64).

[260] Cf. Martens, *Nouveau Recueil Général des Traités*, vol. XVI, pages 345 et seq. On signing and ratifying, Denmark also declared the inclusion of these three straits within the framework of article 35 c), Sweden also declared the same as regards the *Sound*.

4.1.3 Straits Regulated by Specific Treaties Compatible with the Convention Which Are Not Long-Standing

The 'long-standing' requirement in article 35 c) excludes from its scope of application those treaties which regulate passage through a specific strait, however, although they were in force at the time of the conclusion of the 1982 Convention, they had not been in force over a long period of time, but were relatively recent. What happens with these special treaties? Are they applicable counter to the UNCLOS? In the event that there is a treaty with these characteristics, what is their position as regards the provisions of the Convention?

The situation is clearly included in the context of the *application of successive treaties relating to the same subject matter* and, consequently, it will be necessary to retain the regulations in this regard stipulated in article 30 of the *Vienna Convention on the Law of Treaties*, of May 22, 1969.[261] Thus, the first criteria for a solution refers to the existence of a "clause of relation", that is to say, a clause included in one or both successive treaties which expressly regulate their mutual application. In this regard, the UNCLOS includes a 'clause of relation', which is article 311 entitled: "Relation to other conventions and international agreements". The paragraph 2 of this provision is the one which provides the solution to the question posed at this point:

> "2. This Convention shall not alter the rights and obligations of States Parties which arise from other agreements compatible with this Convention and which do not affect the enjoyment of their rights by other party States or the performance of their obligations under this Convention".

In accordance with the stipulations in article 311.2, a treaty which specifically regulates the passage through a determined strait will be applicable on condition that its provisions are compatible with those of the UNCLOS and the rights and obligations in Part III are not affected. That is to say, if the treaty contains a regime of passage for this strait which is more liberal than the one which would correspond by applying the relevant provisions in Part III, then this treaty is applicable.

This is the case of the 1979 Peace Treaty between Egypt and Israel, whose article 5.2 sets out the following:

> "The Parties consider the Strait of Tiran and the Gulf of Aqaba to be international waterways open to all nations for unimpeded and non-suspendible freedom of navigation and overflight. The Parties will respect each other's right to navigation and overflight for access to either country through the Strait of Tiran and the Gulf of Aqaba".[262]

---

[261] A detailed study of these application criteria, as well as their specific application as concerns the UNCLOS, can be seen in López Martín (2002; 2006, p. 241–278).

[262] See the complete text of the Treaty in *ILM*, vol. 18, pages 362–393. For a detailed analysis of this, see Fink (1995, p. 121–144).

There are discrepancies as regards whether the regime stipulated by this provision is similar to that of the freedom of the high seas or to that of passage in transit.[263] No one doubts that, in any case, it is a more liberal regime than the one corresponding to Part III of the Convention.

In fact the Strait of Tiran, a narrow passage 3 miles wide which separates the Gulf of Aqaba from the Red Sea, and is the sole access to the Israeli port of Eilat and to Jordan, belongs to the category of straits included in article 45. 1 b), that is to say, straits located between a part of the high seas or of an Exclusive Economic Zone and the territorial sea of another State; consequently, the applicable legal regime is the right of innocent passage, which is much more restrictive than the one stipulated in article 5.2 of the 1979 Treaty.

Thus, the 1979 Treaty is perfectly compatible with the UNCLOS, in the sense that article 311.2, and is fully applicable as regards passage through the Strait of Tiran in the zone of the territorial sea of Egypt and Israel. We must bear in mind that along the Strait of Tiran there are another two coastal States on its right bank, Saudi Arabia and Jordan, which are not parties to this Treaty. This means that, in the zone of the territorial of these two States, the 1979 Treaty is not in force and the regime of the UNCLOS is, that is to say, the right of innocent passage with no possibility of suspension.

We can conclude by noting that the difference between specific long-standing international conventions on a strait, and other conventions which are also specific as regards a strait in the sense of article 311.2,[264] is that the former may not be compatible with the UNCLOS, a circumstance which may not occur in the latter.[265] Thus, if a long-standing convention stipulates a more restrictive regime of passage than the one which would have been applicable under Part III, the long-standing regime would be applicable, which does not occur with the conventions which are not long-standing.[266] In addition, article 311.2 is applicable not only to conventions previous to the signing of the UNCLOS, but also to international treaties signed previously to the UNCLOS.

---

[263]The different positions in this regard can be consulted in Treves (1985, p. 792).

[264]Logically, the margin of difference stated concerning the fact that it has been in force for some time which is required if it is to be long-standing.

[265]As occurs with the Treaty of April 18, 1990, between Venezuela and Trinidad and Tobago, on the delimitation of marine and submarine areas, whose article VI establishes innocent passage for the Serpent's Mouth Strait and the Dragon's Mouth Strait, where passage in transit must be applied in accordance with article 37 of the UNCLOS. This Treaty is incompatible with the Convention, which gives rise to a problem for Trinidad and Tobago which is a party to the UNCLOS, while Venezuela is not. The text of the 1990 Treaty can be seen in *Law of the Sea Bulletin*, No. 19, 1991, pages 23–27.

[266]Cf. Treves (1985, p. 793).

4.1.4 Straits Through Which a High Seas Route Passes or Which Cross an Exclusive Economic Zone

Article 36 contains an important limitation as regards the applicability of Part III when it excepts:

> "a strait used for international navigation if there exists through the strait a route through the high seas or through an exclusive economic zone of similar convenience with respect to navigational and hydrographical characteristics; in such routes, the other relevant Parts of this Convention, including the provisions regarding the freedoms of navigation and overflight, apply".

Several questions and doubts arise concerning this provision. Thus, it is evident that the provision is intended to indirectly show the interdependence and the necessary confluence of three constitutional components of the concept of 'international strait' in order to fit within the framework of Part III. Considering the geographical and functional components, if the width of the strait is such that it is not covered by the sum of the territorial seas of the coastal States, the legal component is not complied with and Part III is not applicable, not even the right of non-suspendible passage.

Moreover, as T. Treves points out,[267] the second part of the provision is definitely futile as it is evident that, on the high seas and in the exclusive economic zones, neither the right of innocent passage nor the right of passage in transit are applied, while the regulations of Part VII or V are. Nevertheless, having pointed this out, the UNCLOS has bluntly discarded the opinion of a doctrinal sector which claimed the application of the regime on straits for the sea passages which have an uninterrupted band of high sea.[268]

Undoubtedly the most complex is the one concerning the determination that the high sea route or the exclusive economic zone "of similar convenience with respect to navigational and hydrographical characteristics" would be. What does this expression mean exactly? What factors does it depend on? It may be that the 'similitude of convenience' does not depend on the route itself but on the equipment of the vessel so that, for vessels which are perfectly equipped or have sophisticated equipment, the high seas route is just as convenient, however, for other vessels without equipment this is not so and they have to pass through the territorial sea where there are certain navigational aids which makes the transit feasible. More controversially, who decides if this route is equally convenient? The coastal State?, The flag State?, The relevant International Organisation?, A combination of these? According to D. Pharand, this issue can only be resolved through a previously

---

[267]See Treves (1985, p. 790). As stated by this same author, this inclusion was due to a request from ex-Yugoslavia which was trying to safeguard its interests in the Strait of Otranto, between Italy and Albania, due to the pretensions of Albania to establish the extension of its territorial sea beyond 12 miles.

[268]This opinion is held, among others, by Lapidoth (1972, p. 17).

established decision procedure carried out by third parties.[269] Nevertheless, we believe that there is no doubt that this power is attributed to the coastal State and which, in principle, this State must assume this power in accordance with its sovereignty over the waters of the strait. In the event of the inconformity of the States, a controversy will arise and this will have to be resolved through the solution procedures stipulated in Part XV of the UNCLOS, alleging that the coastal State "in exercising the aforementioned freedoms, rights or uses has acted in contravention of this Convention" (article 297 1. b).

Finally, with regard to the stipulations in article 36 of the UNCLOS, an interesting situation may arise. This article does not expressly establish that the *maximum width* which a strait must have in order to apply Part III is *24 miles*. However, this conclusion could be reached by the mere application of logic since, in normal circumstances and, on condition that the particular geography permits it, the coastal States might extend their territorial seas to 12 miles which is the maximum they have a right to under article 5 of the UNCLOS.

However, it may happen that, as we have seen in the previous chapter when we referred to the legal component, the coastal States of a strait decides to establish a territorial sea inferior to 12 miles in such a way that a narrow corridor of high sea or an exclusive economic zone is generated in the centre of the strait. It is perfectly legitimate for a State to limit the claims on its territorial sea. Thus, the coastal State prevents overflight, submerged passage and can suspend innocent passage in its territorial sea, at the cost of ceding part of the width it has a right to in order to create a passage to the high seas.[270]

One example of this was the choice made by Germany as regards the *Kadet*; as opposed to the normal extension of the territorial sea to 12 miles, which, if applied to this sea route, would have meant that it became an international strait, Germany decided to maintain a 3 mile territorial sea in the zone so that a free navigation strip would still exist in the centre of the channel. Moreover, as we mentioned above, Denmark and Sweden attributed territorial seas of 3 miles in order to provide freedom of passage through the *Kattegat* and the *Bornholm*. Japan and Korea acted likewise as regards the two channels of the *Korea Strait*, the *Wester Chosen* and the *Tsushima*; in both zones the two States have territorial seas of 3 miles. Likewise Japan has 3 mile territorial seas as regards the Straits of *Tsugaru-Kaikyo, Osumi-Kaikyo* and *La Pérousse or Soya*, which continue to maintain a high seas route.

As regards this hypothesis, H. Caminos sustains that it can be presumed that the narrow high seas corridor or the exclusive economic zone beyond the territorial sea, is not "of similar convenience with respect to navigational and hydrographical characteristics", therefore, the regime of passage in transit would apply in the straits

[269]See Pharand (1977, p. 76).

[270]Cf. Schachte and Bernhardt (1993, p. 537).

and the right of innocent passage in the areas of the territorial sea.[271] The solution proposed is not very clear.

In our opinion, in the event that the high seas route or the exclusive economic zone between the strips of territorial sea of the coastal States of the strait are not "of similar convenience with respect to navigational and hydrographical characteristics", that is to say, they are not perfectly navigable, it would be necessary to establish the navigable channel in this geographical strait, through the high seas corridor or the exclusive economic zone or in the territorial sea, and only the navigable channel would be considered to be an 'international strait' and Part III would apply to it. The rest of the strait would only be a geographical strait and, therefore, the relevant regime would apply to the strip of high sea or exclusive economic zone and the corresponding regulation would apply to the territorial sea.

An identical solution must be applied in those exceptional cases where the width of the strait is over 24 miles but the strip of high sea or exclusive economic zone which extends beyond the territorial sea of the coastal States is not completely navigable, and the navigable channel penetrates into the territorial sea of one of the coastal States of the strait, only the navigable channel would be considered to be a legal strait. This situation arose in the *Torres Strait*, between Australia and Papua New Guinea until Australia increased the territorial sea of its islands in the strait to 12 miles, which converted it into an international strait. Previously there had been a high seas corridor but this was not navigable while the only channel which is perfectly navigable in the strait lies in the territorial sea of Australia.

Evidently, in the event that the high seas route or the exclusive economic zone is advisable due to the hydrographical and navigation characteristics, that is to say, in the event that there is a navigable channel or an exclusive economic zone in the strait, this would not be a legal strait and would not come under Part III.

In this regard, E. Brüel points out that the situation in which a strip of high seas runs through a strait "rather belongs to the literature than to reality".[272] The Straits of *Soya*, *Tsugaru-Kaikyo*, *Osumi-Kaikyo*, *Western Chosen*, *Tsushima*, *Bornholm*, *Kattegat*, and the *Kadet Channel* already referred to do not appear to be literature or science fiction, they are undoubtedly real.

In any case, the key to this hypothesis is to determine whether the high seas route or the exclusive economic zone are equally advisable as regards the hydrographical and navigation characteristics. Thus, we have an evident problem of interpretation and classification which will have to be resolved when the case arises as the UNCLOS is silent on this matter.

Obviously, there is no doubt that article 36 was drafted thinking of the straits which are over 24 miles wide although the eight straits referred to, which have a

[271]Cf. Caminos (1987, p. 137).

[272]See Brüel (1947, p. 41).

width which is considerably less, also come under its scope of application. The following straits would be included, *a priori*, in article 36.

- *The Skagerrak*. 61 nautical miles wide. Between the north of the Danish Peninsula of Jutland and Norway. It joins the North Sea with the Baltic Sea.
- *The Kattegat*. 12 miles wide and 125 miles long. Its depth varies between 17 and 124 m. It is a continuation of the *Skagerrak* and is located between Denmark (the eastern side of Jutland) and the west coast of Sweden.
- *Bornholm*. 19 miles wide. Between the Danish Island of Bornholm and the south of Sweden, in the Baltic Sea.
- *The Kadet Channel*. 17 miles wide. It joins the Baltic Sea with the Bay of Mecklenburg in Germany.
- Saint George's Channel. 42 miles wide. Between Ireland and Wales (United Kingdom), it connects the Celtic Sea with the Irish Sea.
- *Strait of Sicily*. 53 miles wide. In Italy, between Pallenteria and Sicily, in the Tyrrhenian Sea.
- The *Corsica Channel*. 27 miles wide. Between the Islands of Corsica (France) and Elba (Italy). It joins the Ligurian Sea to the Tyrrhenian Sea.
- The *Malta Channel*. 44 miles wide. Between Italy (Sicily) and Malta, in the Mediterranean Sea.
- Strait of Otranto. 41 miles wide and 40 miles long. Its depth varies between 88 and 732 m. It lies between Albania and Italy and joins the Adriatic Sea and the Ionic Sea.
- Kasos Strait. 27 miles wide. It lies between the Greek Islands of Kasos and Crete.
- Mozambique Channel. 70 miles wide. It lies between Madagascar and Mozambique, in the Indian Ocean.
- Strait of *Vil'kitskii*.[273] 30 miles wide. It is in Russia and joins the Kara Sea to the Laptev Sea.
- *Dimitry Laptev*. 29 miles wide. It is in Russia and joins the Laptev Sea to the East Siberian Sea.
- *Proliv Long*. 75 miles wide. It is in Russia and joins the East Siberian Sea to the Chukchi Sea.
- *Kamchatskiy Proliv*. 93 miles wide. It is in Russia, between the Kamchtka Peninsula and the Island of Nikol'skoje. It joins the Bering Sea to the Pacific Ocean.

[273]It is important to point out that these three Russian Arctic Straits are part of the *Northeast Passage* (Vilkits, Dimitry Laptev and Proliv Long), and have been included among its internal waters by Russia, as it claims these are historic waters, a claim which has no legal grounds. If this Russian vindication fails to prosper, taking into account their width which exceeds double the width of the territorial sea, it is evident that they are excluded from the exclusive sovereignty of Russia, which justifies their inclusion in article 36 of the UNCLOS.

- *Chetvertyy Kuril'skiy Proliv*. 31 miles wide. It is in Russia, between the Kuril Islands of Paramusir and Onekotan. It joins the Ojotsk Sea and the Pacific Ocean.
- *Proliv Kruzenshterna*. 36 miles wide. It is in Russia, between the Kuril Islands of Shiashkottan and Raikoko. It joins the Ojotsk Sea and the Pacific Ocean.
- *Proliv Bussol*. 37 miles wide. It is in Russia, between the Kuril Islands of Urup and Simusir. It joins the Ojotsk Sea and the Pacific Ocean.
- *La Perouse* or *Soya Kaikyo Strait*. 23 miles wide, 25 miles long, and between 20 and 40 m deep. Between the south of the Russian Island of Sajalin or Sakhalin and the north of Hokkaido in Japan (3 miles of territorial sea). It joins the Ojotsk Sea and the Sea of Japan.
- *Tsugaru-Kaikyo*. 10 miles wide with a maximum depth of 200 m. In the north of Japan, between Hokkaido and Honshu (3 miles of territoral sea). It joins the Sea of Japan and the Pacific Ocean.
- *Osumi-Kaikyo* or Van Diemen Strait. 16 miles wide. In Japan, between the Island of Tanega-shima (Ryukyu Islands) and Honshu (3 miles of territorial sea). It joins the East China Sea and the Pacific Ocean.
- *Yakushima-Kaikyo*. 30 miles wide. In Japan, between Yakushima Island and Kuchino-shima Island, in the Ryukyu Islands group. It joins the East China Sea and the Pacific Ocean.
- *Tsushima-Kaikyo* or *East Korea Strait*. 25 miles wide, and approximately 60 miles long. Its depth is approximately 90 m. In Japan, between the Tsushima Island and Iki Island, near the coast of Kyushu (3 miles of territorial sea). It joins the Sea of Japan and the East China Sea.
- *Western Chosen* or *West Korea Strait*. 23 miles wide. Between the Japanese Island of Tsushima and the south of Korea (3 miles of territorial sea). It joins the Sea of Japan and the East China Sea.
- Taiwan or Formosa Strait. 74 miles wide. Between Taiwan and the east coast of China. It joins the East China Sea and the South China Sea.
- *Bashi Channel*, in the *Luzón Strait*.[274] 53 miles wide. Between Taiwan and the Philippine Island of Y'ami. It joins the South China Sea and the Pacific Ocean.
- *Balabac Strait*. 27 miles wide. Between Malaysia and the Philippines (Balabac Island), it connects the South China Sea with the Celebes Sea.
- *Kauai Channel*.[275] 63 miles wide. In the Hawaiian Islands (U.S.A.), between Kauai and Oahu, in the Pacific Ocean.

---

[274] The other two channels which form part of the Luzon Strait (Balitang and Babuyan) are located in archipelagic waters of the Philippines; therefore, they are included in the category of 'archipelagic straits'.

[275] To date, the United States has not made any claims concerning archipelagic delimitation as regards the archipelagos it has sovereignty over, specifically Hawaii and the Aleutians, which means that it maintains the individual delimitation of each of the islands which make up both archipelagos. Neither has the Federal Government recognised any claim concerning archipelagic jurisdiction from any of its States, more specifically, the jurisdiction contained in article XI, section 6 of the constitution of the State of Hawaii, of November 7, 1978, which provides that,

- *Alenuihaha Channel*. 25 miles wide. In the Hawaiian Islands, between Maui and Hawaii, in the Pacific Ocean.
- *Amchitka Passage*. 46 miles wide. In the United States, in the Aleutian Islands, Amchitka Island. It joins the Bering Sea and the Pacific Ocean.
- *Amukta Passage*. 35 miles wide. In the United States, in the Aleutian Islands, West of Amukta Island. It joins the Bering Sea and the Pacific Ocean.
- *Dixon Entrance*. 27 miles wide. Between the Alexander Archipelago in Alaska (United States) and Queen Charlotte Islands (Canada), in the Pacific Ocean.
- *Denmark or Greenland Strait*. 138 miles wide. Between the east coast of Greenland (Denmark) and Iceland. Jan Mayen Island lies to the north east of the strait. It joins the Arctic Ocean and the Atlantic Ocean, and is approximately 300 miles long (480 km).
- *Davis Strait*. 164 miles wide. Between the western coast of Greenland (Denmark) and Baffin Island (Canada). It joins Baffin Bay and the Labrador Sea. Its depth varies between 1,000 and 2,000 m.
- *Lancaster Sound*.[276] 51 miles wide. Between the Canadian Arctic Islands of Bylot and Devon. It joins Baffin Bay and the *Barrow Strait*.
- *Viscount Melville Sound*. 66 miles wide. Between the Canadian Arctic Islands of Melville and Victoria. It joins the *Barrow Strait* and the *McClure Strait*.
- *McClure Strait*. 43 miles wide. Between the Canadian Arctic Islands of Melville and Banks. It joins *Viscount Melville Sound* and the Beaufort Sea.
- *Victoria Strait*. 50 miles wide. Between the Canadian Arctic Islands of Victoria and King William. It joins the *Franklin Strait* and the Queen Maud Gulf.

"The State shall have the power to manage and control the marine seabed and other resources located within the boundaries of the State, including the *archipelagic waters* of the State". Likewise, in the Memorandum of October 24, 1997 of the NOAA, it is expressly stated that "state waters do not include 'channel waters' or 'archipelagic waters" (cf. MacDonald and Mitsuyasu 2000, p. 73 and 81). For the purposes which interest us here, this means that the straits which we cite as included in article 36 as they exceed the 24 miles of territorial sea (Kauai, Alenuihaha, Amchitka, and Amukta), continue to form part of this category.

[276]As regards the eight straits we refer to from Lancaster Sound to and including the Hudson Strait (Viscount Melville Sound, Victoria Strait, Prince Regent Inlet, Foxe Channel, McClure Strait and James Ross Strait), we must add two remarks. The first is to bear in mind the doubts and discrepancies concerning compliance with the functional component we put forward in the previous chapter, and, consequently, the fact that they are straits of an international character.

The second is to take into account that all the straits situated in the Canadian Arctic Archipelago are vindicated by Canada as straits located in its internal waters, and it alleges that they are historic waters. However, as we have pointed out, this claim has no legal grounds nor does it appear to be in accordance with International Law. According to a sector of doctrine, there is the possibility of carrying out delimitation with an archipelagic perimeter in the Canadian archipelago, alleging the "ecological unity"; the ecological characteristics of the arctic marine environment could justify the unification of the waters and lands of this archipelago. In the event that none of these possibilities is backed by the other States, and it seems doubtful that they would receive the backing of the United States, and, taking into account the distance between the islands, the relevant delimitation would be the individual delimitation of each of these, which supposes that the straits mentioned will continue to have a width greater than 24 miles, that is to say, article 36 of the UNCLOS.

- *James Ross Strait*. 30 miles wide. In Canada, between King William Island and the Boothia Peninsula. It joins the *Victoria Strait* and the *Rae Strait*.
- *Prince Regent Inlet*. 40 miles wide. Between the Canadian Arctic Islands of Baffin and Somerset. It joins *Lancaster Sound* and the Boothia Gulf.
- *Foxe Channel*. 80 miles wide. Between the Canadian Arctic Islands of Baffin and Southampton. It joins the *Hudson Strait* and Foxe Basin.
- *Hudson Strait*. 60 miles wide. In Canada, between Baffin Island and the Ungava Peninsula. It joins the *Foxe Channel* and the Labrador Sea.
- *Cabot Strait*.[277] 57 miles wide. In Canada, between the island of Newfoundland and the Canadian mainland. It joins the Atlantic Ocean and the Gulf of St. Lawrence.
- *Honguedo Strait* or *Gaspe Passage*. 38 miles wide. In Canada, between Anticosti Island and the Gaspe Peninsula, in the Gulf of St. Lawrence.
- *Florida Strait*. 82 miles wide and 330 miles long. Its maximum depth is 1,042 m. It is situated between Key West (U.S.A.) and Cuba. It joins the Atlantic Ocean and the Gulf of Mexico, and the Keys of Florida and Cuba.
- *Yucatan Strait or Channel*. 105 miles wide. Between Cuba and the Peninsula of Yucatan (Mexico). It joins the Caribbean Sea and the Gulf of Mexico.
- *Caicos Passage*. 35 miles wide. Between the Island of Mayaguana (Bahamas) and the Caicos Islands (United Kingdom), in the Atlantic Ocean.
- *Windward Passage*. 45 miles wide and with a depth of 1,700 m. It is 40 miles long. It separates Cuba from Haiti. It joins the Atlantic Ocean and the Caribbean Sea, and is the most direct link with the Panama Canal.
- *Jamaica Channel*. 69 miles wide. Between Jamaica and Haiti. It is the continuation of the *Windward Passage*.
- *Mona Passage*. 33 miles wide. Its depth varies between 61 and 274 m, and it is 50 miles long. It is located between the Dominican Republic and Mona Island (Puerto Rico). It joins the Atlantic Ocean and the Caribbean Sea, and is an important navigation route between the Atlantic and the Panama Canal.
- *Anegada Passage*. 48 miles wide. Between Anegada Island and Sombrero Island, in the Virgin Islands (United Kingdom). It joins the Atlantic Ocean and the Caribbean Sea. It has a depth of 7,550 feet (2,300 m), it is the deepest channel in the eastern Caribbean.

---

[277] Both in the cases of the Cabot Strait and the Honguedo Strait, there is a particular legal problem. Both straits, which are the continuation of each other, far exceed 24 miles, therefore in principle they are included in article 36. However, Canada has drawn straight baselines (Order of May 9, 1972) in such a way that both are included in its internal waters. Nevertheless, there are discrepancies concerning whether these baselines are in accordance with the method in article 7, therefore, thee have been protests from several states including the United States. This circumstance casts doubts on this delimitation, therefore, if it is not in accordance with International Law, these straits will continue within the scope of article 36. However, if the delimitation is valid, these two straits would become part of the category included in article 45.1 b), and the regime would be that of the right of non-suspendible passage. Obviously, the difference in legal regime is important.

- *Bransfield Strait*. 60 miles wide, with a length of approximately 300 km lying roughly northeast–southwest. Between the South Shetland Islands in the north and the Antarctic Peninsula in the south, in the Glacial Arctic Ocean. The strait is called the 'Mar de la Flota' by the Argentineans.
- *Drake Passage*. 400 miles wide. Between Cape Horn to the north and the South Shetland Islands to the south. This is the natural separation between South America and the Antarctic.

4.1.5 Straits Which Include Archipelagic Waters as a Consequence of the Drawing of Archipelagic Perimeters in Archipelagic States

The final category of straits which falls outside the scope of application of Part III is composed of the so called 'archipelagic straits', that is to say, the international straits which have been included in the archipelagic waters of an archipelagic State as a result of the drawing of archipelagic perimeters.[278] These straits would cease to be characterised as such depending not only on the very nature of the archipelagic waters (the waters used to measure the territorial sea), but also on the areas joined (areas merely archipelagic as a result of the implementation of the archipelagic principle).

The legal regulation of this category of straits is the one included in Part IV: "Archipelagic States". Although an important part of this regulation is included in Part III as it was passed on to this Part. Specifically, as regards the "duties of ships and aircraft during their passage, research and survey activities, duties of the archipelagic State and laws and regulations of the archipelagic State relating to archipelagic sea lanes", "*mutatis mutandi*" article 54 passes on the provisions in articles 39, 40, 42 and 44 as regards straits.

We should point out that this exclusion does not expressly appear in any provision of Part III, just as there is no article in Part IV which refers specifically to international straits. What is the legal basis for including the archipelagic straits in Part IV? The answer is in the provision in article 53. We understand that the reference in this provision to the "archipelagic sea lanes" clearly includes the international straits included in archipelagic waters as these straits are obviously 'archipelagic sea lanes'. The total consideration of this provision, together with the fact that Part III does not include the straits which connect archipelagic straits

[278] We must point out that the concept of 'archipelagic strait' is restricted only to the straits located in the archipelagic waters of an archipelagic State as regulated by Part IV. By applying strict international legality, we cannot extend the concept of archipelagic strait to the straits located within an archipelagic perimeter drawn in 'State archipelagos', as is the case of Hawaii, since it is commonly acknowledged that Part IV cannot be applied to these archipelagos, which must individually delimit each island. This exclusion of State archipelagos from the drawing of the archipelagic perimeter has received abundant, justified criticism, which we will not go into as it exceeds the limits of this work.

among the categories of straits it applies to justifies the inclusion of archipelagic straits in Part IV. According to Maduro,[279] the reason which explains that article 53 does not use the expression 'straits' is that the waters in question are not territorial seas, but archipelagic waters.

We can list the following international straits which were included in archipelagic waters after the drawing of the corresponding archipelagic perimeter by the States which have claimed the condition of archipelagic State[280] as provided in articles 46 and 47 of the UNCLOS. The navigation regime applied to these will be "the right of archipelagic sea lanes passage" in article 53, similar to passage in transit, when these straits are designated by the coastal State as archipelagic sea lanes, or the right of innocent passage in article 52 (with the possibility of suspension) in the event that they are not designated as such. Although it should be mentioned that, to date, the only archipelagic State which has designated the pertinent archipelagic sea lanes in accordance with article 53, is Indonesia. As a consequence of these drawings, the Straits of *Karimata, Sunda, Makassar, Lombok, Leti and Ombai* were designated as archipelagic sea lanes, therefore, the passage is similar to that of transit; all the other straits of Indonesia will be subject to the right of innocent passage with the possibility of suspension.[281]

- In **The Philippines**:[282]
  - *Balintang Channel.* This is the smallest sea lane which separates the Batanes Islands and the Babuyan Islands in the *Luzon Strait.*
  - *Babuyan Channel.* 15 miles wide. Between the Babuyan Islands and Luzon.
  - *Polillo Strait.* 10 miles wide. Between the Polillo Island and Luzon.
  - *San Bernardino Strait.* This has a maximum width of 24 miles and a minimum of 2. It is funnel shaped, and is wider in the east than in the west. It has a length of approximately 35 miles, and its depth varies between 30 and over 100 fathoms. It separates the Bicol Peninsula in the island of Luzon from Samar Island to the south. This is the most easterly strait of the Philippines.

---

[279] Cf. Maduro (1980, p. 88).

[280] The archipelagic States which have claimed the condition of archipelago and there are international straits in their archipelagic waters are the following: The Bahamas (Law No. 37 of 1993), Fiji (Order No. 119 of 1981), the Philippines (Law No. 5446 of 1968), Indonesia (Law No. 6 of 1996 modifying the Law of 1960, and Regulation No. 61 of 1998 on coordinates), The Solomon Islands (Law No. 32 of 1978 and Legal Notice No. 41 of 1979 on archipelagic baselines), Papua New Guinea (Declaration of July 25, 2002), and Trinidad and Tobago (Law No. 24 of 1986 and Order No. 206 of 1988 on archipelagic baselines). The references can be seen at http://www.un.org/Depts/los/LEGISLATIONANDTREATIES/regionlist.htm.

[281] The description of these sea lanes, as well as the corresponding map, found in the *Law of the Sea Bulletin*, No. 52, 2003, pages 21–41.

[282] The Philippines has claimed the classification of internal waters for the waters included within its archipelagic perimeter, and this is included in its Law of September 18, 1968. However, its ratification of the UNCLOS of July 23, 1997 obliged it to relinquish this claim and it must modify its internal legislation, which it has not yet done.

- *Maqueda Channel*. 4 miles wide. Between the Islands of Cataduanes and Luzon.
- *Verde Island Passage*. 4 miles wide. Between the Islands of Luzon and Mindoro, with Verde Island in the middle. The distance between Verde Island and Luzon is 3 miles.
- *Mindoro Strait*. 20 miles wide. Between the Calamian Islands and Mindoro Island (the distance between Mindoro and Apo Reef is 15 miles).
- *Suriago Strait*. Its width varies from 8.5 to 26 miles. It is hook shaped and is 70 miles long. It is quite deep: in the northern half it varies between 18 and 60 fathoms, while the southern half goes as deep as 770 fathoms. It lies between Leyte and Mindanao and connects the Bohol Sea and the Leyte Gulf.
- *Basilan Passage*. 7 miles wide. Between Mindanao and the Sulu Archipelago.
- *San Juanico Strait*. 1 mile wide. Between the Samar Islands and Leyte.
- *Bohol* or *Cebu Strait*. This separates the Islands of Cebu and Bohol, and connects the western part of the Bohol Sea and the Camotes Sea.
- *Tanon Strait*. This lies between the Islands of Negros and Cebu. It connects the Visayan Sea with the Bohol Sea.
- *Sibutu Passage*. 18 miles wide. Located in the Sulu archipelago near Borneo.

• In **Indonesia**:
- *Selat Lombok*. This has a minimum width of 11 miles and a maximum of 32 miles. It joins the Java Sea and the Indian Ocean and lies between the Islands of Bali and Lombok.
- *Makassar Strait*. 62 miles wide, between Kalimantan (Borneo) and the Island of Sulawesi (Celebes). This is complementary to the *Lombok Strait*. Its southern entrance is divided into three channels as a result of the existence of coral reefs. It joins the Celebes Sea and the Flores Sea.
- *Selat Serasan*. 23 miles wide. On the northeast coast of Borneo.
- *Selat Berhala*. 9 miles wide. Between Singkep and Sumatra. From Berhala to Singkep there is a distance of 10 miles.
- *Selat Bangka*. 8 miles wide. Between Bangka and Sumatra.
- *Selat Gaspar*. 8 miles wide. Between Bangka and Billiton.
- *Selat Sunda*. This has a width of 31 miles maximum and 5 miles minimum (at Pulau Sangiang). Its depth goes from 16 to more than 60. It is located between Java and Sumatra. It connects the Java Sea and the Indian Ocean. Its northern entrance is located approximately 50 miles west of Djakarta. There are small islands all along the strait.
- *Selat Gelasa*. 40 miles wide. Between Pulau Bangka and Belitung.
- *Selat Mentawai*. This is approximately 45 miles wide. It lies between Sumatra and Kepulauan.
- *Selat Karimata*. This is 66 miles wide. It lies between Sumatra and Borneo, south of Kalimatan Island. It connects the South China Sea with the Java Sea.
- *Selat Bali*. 2 miles wide. It lies between Bali and Java, at Nusa Tenggara.
- *Selat Badung*. 13 miles wide, and approximately 60 km long. It lies between Bali and Nusa Penida.

- *Selat Alas.* 5 miles wide. It lies between Lombok and Sumbawa, at Nusa Tenggara.
- *Selat Sape.* 8 miles wide. It lies between Komodo and Sumbawa.
- *Selat Grehund.* 10 miles wide. Near the Celebes Islands to the east.
- *Koti Passage.* 10 miles wide. Near the northeast coast of Borneo.
- *Api passage.* 16 miles wide. Near the northeast coast of Borneo.
- *Selat Timpaus.* 10 miles wide. Between Banggai and Sula (Celebes Islands). It joins the Molucca Sea and the Banda Sea.
- *Selat Manipa.* 12 miles wide. Between Buru and Ceram. It joins the Ceram Sea and the Banda Sea.
- *Selat Leti.* It joins the Timor Sea and the Banda Sea.
- *Selat Roti.* 6 miles wide. Between Roti and Timor.
- *Selat Ombai.* This has a width which varies between a maximum of 32 miles and a minimum of 13. It separates the Alor Archipelago from the Wetar, Atauro and Timor Islands, in the Lesser Sunda Islands. It joins the Banda Sea in the north with the Savu Sea to the southeast. Its continuation is the *Wetar Strait*, between Indonesia and East Timor.[283]

• In **Papua New Guinea**:
  - *Isumru*d or *Goschen Strait.* 7 miles wide. Between New Guinea and the D'Entrecasteaux Islands.
  - *Vitiaz Strait.* 24 miles wide. Between New Guinea and the Bismarck Archipelago.
  - *St. George's Channel.* 8 miles wide. Between New Britain and New Ireland.

• In the **Solomon Islands**:
  - *Bougainville Strait.* 15 miles wide. Between Boungainville and Choiseul.
  - *Manning Strait.* 6 miles wide. Between Choiseul and St. Elizabeth.
  - *Indispensable Strait.* 19 miles wide. Between Guadalcanal and Malaita.

• In **Fiji**:
  - *Vatu-i-Ra Channel.* 2 miles wide at its narrowest part. It lies between Viti Levu and Vahua Levu, in the Pacific Ocean.

• In the **Bahamas**:
  - *Northeast Providence Channel.* 29 miles wide. Located between Great Abaco Island and Eleuthera Island.
  - *Northwest Providence Channel.* 26 miles wide. Located to the southwest of Great Abaco Island.
  - *Crooked Island Passage.* 26 miles wide, between Long island and Crooked Island.
  - *Mayaguana Passage.* 39 miles wide, between Acklins Island and Mayaguana Island.

• In **Trinidad and Tobago**:
  - *Galleons Passage.* 19 miles wide, between Trinidad and Tobago.

[283] It should be remembered that, of all these straits we have mentioned, only the straits of Karimata, Sunda, Makassar, Lombok, Leti and Ombai are archipelagic sea lanes.

4.2 Straits in Which the Regime of Navigation Is the Right of Innocent Passage

There are two categories of straits used for international navigation with the regime of the right to innocent passage with no possibility of suspension, which operate as exceptions to passage in transit which was so laboriously claimed at the III Conference by the Great Maritime Powers, as we have seen. These cases are expressly included in articles 38.1 and 45.1 b), respectively.

4.2.1 Straits Formed by an Island of a Coastal State Bordering the Strait and Its Mainland Territory

Article 38.1 UNCLOS excludes the application of the right of passage in transit:

> "if the strait is formed by an island of a State bordering the strait and its mainland, transit passage shall not apply if there exists seaward of the island a route through the high seas or through an exclusive economic zone of similar convenience with respect to navigational and hydrographical characteristics".

As regards these types of straits, the right of innocent passage is applied, as provided in article 45.1 a): "The regime of innocent passage... shall apply in straits used for international navigation: a) excluded from the application of the regime of transit passage under article 38, paragraph 1".

This exception is commonly known as the "Messina Clause", as it was expressly included in order to be applied to the *Strait of Messina*, which separates the Italian Island of Sicily from the mainland territory of Italy.[284] This strait can be taken as a reference point for the application of this exception and as regards the interpretation of what could be a similar convenience route as the vessels which avoid the *Strait of Messina* will travel by a route which is 60 miles longer than the route through the strait.

However, the text of article 38.1 is confuse, imprecise and unsatisfactory, and leaves many controversial questions unanswered.

In the first place, we must relate this provision with article 36, already analysed above, as regards the existence of "a route through the high seas or through an

---

[284] In the course of the III Conference, Italy began to worry about the pressure of the Great Maritime Powers concerning the imposition of freedom of transit in all the international straits as this entailed important strategic implications if 'enemy' or 'unfriendly' powers chose to use the navigation route through the Strait of Messina. Taking into account that there is a perfectly feasible alternative high sea route to the southeast of Sicily, Italy argued that vessels could choose this other free navigation route, and for navigation purposes, the waters of these types of straits must continue to have the same status as that of the territorial sea, that is to say, they should continue to be governed by the right of innocent passage. This was how it was included in its proposal for articles (cf. UN Doc. A/AC.138/SC.II/L.30 (1973).

exclusive economic zone of similar convenience with respect to navigational and hydrographical characteristics", bearing in mind that both articles use an identical expression. This means that the same questions we referred to above concerning the meaning of these terms, the factors to be taken into consideration in order to establish convenience, or the competence for the decision on the existence or not of an alternative route, and in this regard we refer to what we have stated above. However, following the opinion of K.L., Koh[285] we believe that it is important to point out that convenience only refers to hydrographic and navigation characteristics, therefore, economic, military and political convenience is irrelevant. This means that the States which use the straits cannot invoke these factors in order to exclude a strait from the exception included in article 38.1.

The question concerning whether this exception is only applied to an island and its mainland or whether it is possible to apply this to the case of a strait which separates two islands of the same State is also more controversial when there is an equally convenient high sea route or exclusive economic zone on the other side of these islands. This would be the case as regards the *Menorca Channel* which separates the Islands of Menorca and Mallorca in Spain and has other navigation routes. A literal interpretation of article 38.1 leads to a clearly negative response. This constitutes another proof of the lack of rigour of some of the provisions of the Convention concerning straits. Another case would be that of two islands, one of which is considered with certainty to be the "mainland" of an island State – undoubtedly the larger. In this case, if there is an alternative high sea route or an exclusive economic zone on the other side of the smaller island (the one which is not considered to be the mainland), this is clearly included in article 38.1; among other cases, this includes *The Solent*, Japan with *Sado-Kaikyo*, and New Zealand with the *Foveaux Strait*.

As regards this problem, it should also be mentioned that there is a specific situation which arises in the case of the existence of several islands belonging to the same State, where there are a number of international straits. Should be included in article 38.1 and the rest in article 37, that is to say, in one of the many straits innocent passage would be in force and in transit passage in the others? The difficulty of this option is determining which straits would come under which category. Besides this, who would take this decision? This is so complex that, in the opinion of T. Treves, this hypothesis is excluded from the scope of application of article 38.1.[286]

This particular case occurs in Greece, with the numerous islands (Cyclades and Dodecanese, for example) it has in the Aegean Sea. In this regard, Greece signed and ratified the UNCLOS, which showed that, in these cases, the coastal State has the competence to designate the route or routes where passage in transit will be in force. This opinion is shared by the former Yugoslavia, subsequently by Serbia, and more recently by Montenegro, which have also declared their competence to

[285] Vid. Koh (1982, p. 150).

[286] Cf. Treves (1985, p. 790).

designate in which straits of article 38.1 innocent passage is in force; however, this was categorically opposed by Turkey, which considered that this was creating a new category of international straits contrary to the Convention.[287] Again it is certain that the literal meaning of the UNCLOS does not seem to permit this possibility.

These specific situations we have referred to and which, illogically and incomprehensibly, do not come under the literal terms of article 38.1, lead us to claim that this exception must not be interpreted mechanically, but should be applied with common sense, taking the geographical factors and other relevant circumstances into account.[288]

Besides this pretension, the straits which clearly come under the framework of article 38.1 would be the following:[289]

- *The Solent.* 2 miles wide. Between the Isle of Wight and the mainland of the United Kingdom, in the English Channel.
- *Pentland Firth.* 5 miles wide. Between the Orkney Islands and Scotland, it joins the Atlantic Ocean and the North Sea.
- *North Minch.* 20 miles wide. Between the Outer Hebrides and Scotland (United Kingdom), in the Atlantic Ocean. It continues through the *Little Minch* which, as it is situated between two islands and separates two exclusive economic zones, is no longer under article 38.1, and now falls under article 37.
- *Kalmarsund* or *Kalmar Strait.* 2 miles wide. Between Öland Island and the southeast mainland of Sweden, in the Baltic Sea.
- *Kildin Strait.* Less than a mile wide and 10 miles long. Between Kildin Island and the north coast of Russia (in Murmansk), in the Barents Sea.
- *Elba Strait* or *Piombino Channel.* 5 miles wide. Between the Island of Elba and the western mainland of Italy. It joins the Tyrrhenian Sea and the Ligurian Sea.

---

[287] The text of the respective declarations and objections can be seen in http://untreaty.un.org. We should point out that Turkey has not signed nor is it a party to the UNCLOS.

[288] Cf. Nandan and Rosenne (1993, p. 329).

[289] We should point out that, in the north of Canada, between its coast and the Islands of the Arctic archipelago, there are several straits which comply with the conditions of article 38.1. However, these straits are not used for international navigation, therefore in principle they would not come under Part III. These are the following straits: *Franklin Strait*: 20 miles wide, between Prince of Wales Island and the Boothia Peninsula; *Dease Strait*: 12 miles wide, between Victoria Island and the Kent Peninsula; *Rae Strait*: 13 miles wide, between King Williams Island and the Gibson Peninsula; *Dolphyn and Union Strait*: 15 miles wide, between Victoria Island and the north of the Canadian mainland; *Simpson Strait*: 2 miles wide, between King Williams Island and the Adelaide Peninsula; *Bellot Strait*: One and a half miles wide, between Somerset Island and the Boothia Peninsula; *Fury and Hecla Strait*: half a mile wide at its narrowest part, between Baffin Island and the Melville Peninsula. In fact, as we said in the previous chapter, the advisable route through the *Northwest Passage* and the one occasionally used is the so-called Route 1, which constitutes the alternative free navigation route which exists at the other side of the aforementioned straits.

- *Strait of Messina*. Its width varies from less than two miles (between Faro Point and Rock of Scylla) and 10 miles (between Capes Ali and Pellaro). It is 30 miles long. It separates the Island of Sicily to the west and the Italian mainland to the east. It's linking the Tyrrhenian and Ionian Seas.
- *Corinth Channel*. This is less than one mile wide. It separates the Peloponnesians from the southeast mainland of Greece. It joins the Gulf of Corinth and the Saronic Gulf.
- *Euripus Strait*. Approximately one mile wide. Between Euboea Island and mainland Greece to the west, in the Aegean Sea.
- *Corfu Strait*. 3 miles wide. Between the Greek Island of Corfu and the Greek and Albanian mainlands. It linking the Ionic and Adriatic Seas.
- *Nikola Strait*. Between the island of Sveti Nikola and the mainland of Montenegro, opposite the city of Budva, in the Adriatic Sea.
- *Mijetski Kanal*. One and a half miles wide. Between Otok Mljet and the Croatian Peljesac Peninsula in the Adriatic Sea.
- *Pemba Channel*. Between 30 and 20 miles wide, and approximately 60 miles long. Between the Tanzanian Island of Pemba and the eastern mainland of Kenya and Tanzania, in the Indian Ocean.
- *Zanzibar Channel*. This has a minimum width of sixteen miles and a maximum width of twenty-one and a half miles. Its length is approximately 80 miles, and its depth varies between 10 and 40 fathoms. It is the continuation of the Pemba Channel. It is situated between the Island of Zanzibar and the southeast coast of the Tanzanian mainland.
- *Proliv Litke*. 13 miles wide. It separates the eastern Island of Karaginsky from the Russian Kamchatka Peninsula, in the Bering Sea.
- *Tartaria Strait* or *Tatarskiy Proliv* and *Proliv Nevel'skogo*. These constitute two independent straits, however, as each one is the continuation of the other, they must be considered as a whole. They separate the Russian Island of Sakhalin from southeast Russia in mainland Asia. It is 900 km long (486 miles), and connects the Ojotsk Sea and the Sea of Japan.
  The *Tartaria Strait* is approximately 28 miles wide, and has a maximum depth of 20 m. While the width of the *Proliv Nevel'skogo* is no greater than 10 miles.
- *Rishiri-Suido*. 10 miles wide. Between Rishiri Island and the Wakku Peninsula in Hokkaido (Japan), in the Sea of Japan.
- *Sado-Kaikyo*. 18 miles wide. Between Ryotsu Island and the west coast of Honshu (Japan), in the Sea of Japan.
- *Cheju Haehyop* or *Cheju Strait*. 12 miles wide. Between Cheju Island and the southwest mainland of South Korea. It joins the Sea of Japan and the Yellow Sea.
- *Pescadores Channel* or *P'eng-hu Shuitao or P'eng-hu Kang-tao*. 17 miles wide. Between Pescadores Island to the west and mainland Taiwan to the east. It connects the East China Sea and the South China Sea.

- *Hainan* or *Quiongzhou Strait*. 10 miles wide, and with a length of 40 miles between its natural entrance points. Its depth varies between 17 and 40 fathoms. It is located between Hainan Island and mainland China, in the Gulf of Tonkin in the South China Sea.
- *Colville Channel*.[290] This is located between Great Barrier Island and the Coromandel Peninsula in North Island (New Zealand), in the Pacific Ocean. It forms the entrance to the Hauraki Gulf.
- *Foveaux or Floveaux Strait*. 15 miles wide and 45 miles long. Its depth varies between 12 and 25 fathoms. It is located between Stewart al Island to the south and South Island (New Zealand) to the north. It joins the Pacific Ocean and the Tasman Sea.
- *Etolin Strait*. 17 miles wide and approximately 60 miles long. Between Nunivak Island and the southwest mainland of Alaska (U.S.A.), in the Bering Sea.
- *Shelikof Strait*. 20 miles wide. Between Kodiak Island and the southeast mainland of Alaska (U.S.A.), in the Pacific Ocean.
- *Icy Strait*. One mile wide and 40 miles long. Between Chicago island and the south of Alaska (U.S.A.).
- *Hecate Strait*. 24 miles wide. Between Queen Charlotte Islands and the western mainland of Canada, in the Pacific Ocean.
- *Queen Charlotte Strait and Georgia Strait*. The width varies from 35 to 15 miles and the length is approximately 200 miles. They are the continuation of each other.[291] They are located between Vancouver Island and the western mainland of Canada (British Columbia), in the Pacific Ocean.
- *Santa Barbara Channel*. 11 miles wide and approximately 80 miles long. Between the Channel Islands and the coast of California (U.S.A.), in the Pacific Ocean.
- *Strait of Belle Isle*. 9 miles wide and approximately 125 km (67 miles) long. Between the Island of Newfoundland and the Labrador Peninsula (Canada). It joins the Atlantic Ocean and the Gulf of Saint Lawrence. The name derives from a small island called Belle Isle, which is located at the eastern end of the strait.
- *Jacques Cartier Strait*. 15 miles wide. Between Anticosti Island to the south and the Canadian mainland of Quebec to the north, in the Gulf of Saint Lawrence.

---

[290]This strait links the *Cradock Channel* and the *Jellicoe Channel*, which joins the internal waters of New Zealand and the Pacific Ocean in the Hauraki Gulf.

[291]The *Georgia Strait* finally links with the *Strait of Juan de Fuca*, on the border with the U.S.A. We must point out that the *Queen Charlotte* and *Georgia Straits* join with other minor straits which run between the coast of Canada and Vancouver Island; specifically *Johnston Strait* (from Thurston Bay to Mamalilaculla) and *Broughton Strait* (in the Broughton Archipelago).

- *Northumberland Strait.* 7 miles wide. Between Prince Edward Island to the north and the Nova Scotia Peninsula (Canada) to the south, in the Gulf of Saint Lawrence.
- *Le Maire Strait.* 16 miles wide. Between Tierra del Fuego and Estados Island (Argentina), in the Atlantic Ocean.

4.2.2 Straits Located Between One Part of the High Seas or an Exclusive Economic Zone and the Territorial Sea of Another State

Article 45 of the UNCLOS stipulates the following:

> "1. The regime of innocent passage, in accordance with Part II, section 3, shall apply in straits used for international navigation:
>
> b) between a part of the high seas or an exclusive economic zone and the territorial sea of a foreign State".

This second exception to passage in transit corresponds to a category already included in article 16.4 of the 1958 Geneva Convention; therefore, it maintains the legal regime codified in this article concerning non-suspendible innocent passage.

This provision incorporates the proposal presented by the United Kingdom and that of the USSR and others, which clearly distanced itself from the proposal of the U.S.A. which proposed the application of passage in transit in all the straits, even those which joined territorial seas. In this regard, the representative of the British delegation justified this exception by alleging that,

> "with regard to straits used for international navigation between one part of the high seas and the territorial sea of a foreign State, the interest of the International Community in free navigation is not so strong as in the case of straits linking two parts of the high seas".[292]

This approach has led to classifying these types of straits as "secondary", in contrast with those which join two parts of the high seas or exclusive economic zones, which would be "principal straits".

As pointed out by Schachte and Bernhardt,[293] article 45.1 b) is intended for the "dead-end" straits, that is to say, those which do not have an outlet and finish in the territorial sea of another State which is not one of the coastal states of the strait.

---

[292]See A/CONF.62/C.2/L.3. As regards this particular aspect, it is interesting to point out the different position of the United Kingdom and the U.S.A., two States which belong to the NATO; however, they have different strategic interests, as is clear from this point.

[293]See Schachte and Bernhardt (1993, p. 54).

Thus, this provision has the advantage that it creates a right of passage in the territorial sea of the other State and its neighbours.

Applying the strict literalness of article 45.1 b), we only find three straits which comply with the parameters set out in this provision, in the sense of a strait located between the high seas or an exclusive economic zone and the territorial sea of "a foreign State" ("otro Estado" in the Spanish text, and "autre État" in the French text) which is not a coastal State of the strait.

- This is the case of the Strait of *Tiran*.[294] A narrow corridor which is approximately a minimum of 3 miles, and a maximum of 17, with an approximate length of 98 miles (150 km) and a depth of which varies from 500 to 1,000 m. It is located in the Sinai Peninsula (Egypt) and the west coast of Saudi Arabia and joins the Red Sea and the Gulf of Aqaba. It is the only access to the Israeli port of Eilat and the Jordanian port of Aqaba. The entrance to the Gulf is complicated by the islands which divide the strait into two channels: *Enterprise Passage* (between Sinai and Tiran Island), which is the only navigable channel of the *Strait of Tiran* (from which it receives its name), and the *Graftu Passage* (between Saudi Arabia and Sanafir Island).
- Also included within the framework of article 45.1 b) is the *Väinameri*,[295] a system of straits which connect the Baltic Sea and the territorial sea of Latvia in the Gulf of Riga, and which run between the Estonian Islands of Saaremaa, Hiiumaa, Muhu and Vormsi and the western mainland of Estonia. Two of these straits, which are a continuation of each other, are of interest for navigation.
- The *Hari or Kurk Strait*, which has a width of between 8 and 10 km (4.3 and 5.4 miles), and a depth of approximately 10 m. It separates the Estonian Islands of Hiiumaa and Vormsi. It is the most northerly strait in the Baltic Sea and joins the *Suur Strait*, as far as the Gulf of Riga.
- The *Suur Strait* is narrower than the *Hari Strait* as its width is no greater than 4 km (a little over 2 miles). Its maximum depth is 20 m and the minimum depth 5 m on the southern side. It runs between Saaremaa Island at west, and the western mainland of Estonia to the east. It links Baltic Sea – across *Hari Strait* – with the territorial sea of Latvia in the Gulf of Riga. In fact, it is one of the two entrances to the Latvian port of Rigas Jurmala located in this gulf.[296]
- Finally, we must cite the *Entrance to the Gulf of Finland*, which is 17 miles wide. It is located between Estonia to the south and Finland to the north, and

[294]As we stated in Sect. 4.1.3 in this chapter, in the *Strait of Tiran* there is a dual regime concerning passage. The freedom of transit established in the Egyptian-Israeli Treaty of 1979 is applied in the territorial sea of Egypt, while, in the territorial sea of Saudi Arabia (which is not a party to this treaty), article 45.1 b) of the UNCLOS must be applied, that is to say, the right of non-suspendible innocent passage.

[295]Väinameri means "sea of straits" in Estonian.

[296]The other is the *Irben Strait*, between the south of the Estonian Island of Saaremaa and the northern mainland of Latvia.

connects the Baltic Sea and the Gulf of Finland and the territorial sea of Russia. It is the only entrance to the Russian port of Saint Petersburg.

- However, if we carry out an extensive rather than a strict interpretation of this rule, we could also consider that the straits located between the high sea or an exclusive economic zone and the territorial sea of a State, even if it is a coastal State of the strait are also included. This interpretation would be founded on two considerations. On the one hand, the stipulations in article 35 a) concerning internal waters which we have analysed; on the other hand, the types of straits we refer to are clearly "dead-end" straits, a category which is unanimously considered to be the objective of article 45.1 b), as was stated above. In addition, this same interpretation would also be supported by the opinion of some States.[297] As regards this point, on presenting the proposal of the United Kingdom to the Second Committee, which is the proposal of article 45, the British delegate referred to these types of straits as "linking a part of the high seas with the territorial sea of *a State*".[298]
- The following straits would then be governed by the right of non-suspendible innocent passage.

- *Irben Strait*. Fourteen and a half miles wide, and with a depth of approximately 21 m. It is located between Saaremaa Island (Estonia) to the north, and the northern mainland of Latvia to the south. It joins the Baltic Sea and the Gulf of Riga, which includes territorial waters of both Estonia and Latvia. It is also one of the entrances to the port of Riga.
- *Wetar Strait*. It's located between Pulau Wetar (Indonesia to the north) and East Timor to the south. It is 19 miles wide at its narrowest point. It joins the Arafura Sea and the archipelagic waters of Indonesia.[299]
- *Strait of Juan de Fuca*. 9 miles wide and 70 miles long. At some points it is more than 100 fathoms deep. It separates the south of Vancouver Island (Canada) to the north and the Olympic Peninsula in Washington (U.S.A.) to the south. It connects the Pacific Ocean and the territorial waters of Canada and the U.S.A. It continues to the west through the Georgia Strait and to the east through the Puget Sound.
- *Head Harbour Passage*. between the Peninsula of Nova Scotia (Canada) and the coast of Maine (U.S.A.). It connects the Atlantic Ocean and Passamaquoddy Bay in the U.S.A.

---

[297]This would be the case of the U.S.A., which considers that Head Harbour Passage which leads to Passamaquoddy Bay (U.S.A.) crossing the territorial sea of Canada is included in article 45 (cf. *Limits in the Seas*, No. 112, p. 66).

[298]See Second Committee, 11th meeting (1974), p. 24, *II Off. Rec.* 125 (the bold lettering is ours).

[299]As we mentioned at the beginning of the second heading of this chapter, the *Wetar Strait* is in a special category of strait which does not respond to any of those stipulated in the Convention. We place it in the right of innocent passage by establishing an analogy between archipelagic waters and territorial for the purposes of passage.

If this extensive interpretation is not admitted, these straits would not be included in Part III, but in Part II of the UNCLOS, and its navigation regime would be the right of innocent passage with no possibility of suspension.

4.3 The "Principal" States or States Subject to the Right of Passage in Transit

After all the pertinent exceptions to the Part III regime in general and to the right of transit in particular come into operation, it is finally necessary to address the question of the "normal" straits, that is to say, the category of straits governed by passage in transit. This is the most numerous category and includes the most important straits for international navigation, from strategic, military, economic and commercial points of view. Such circumstances justify the denomination of this particular category of straits as "principal".

These are referred to in article 37 of the UNCLOS, which determines the scope of application of the Second Section of Part III concerning passage in transit, in the following terms:

> "This section applies to straits which are used for international navigation between one part of the high seas or an exclusive zone and another part of the high seas or an exclusive economic zone".

These are straits which join two free navigation zones; and this has been the "traditional" category of international straits for excellence and codified in article 16.4 of the 1958 Geneva Convention. For obvious reasons, it was completed with the new maritime area, the exclusive economic zone. As we stated above, these are the "traditional" straits of International Law, and, despite this or perhaps due to this, these are the straits affected by the new legal regime included in the UNCLOS: the important right of passage in transit, the compromise regime which arose from the III Conference as a result of pressure from the great maritime powers.

These "normal" or "principal" straits[300] are those listed below:

- *The Hole*. 23 miles wide. Between the Orkney Islands and the Shetland Islands (United Kingdom). It joins the Atlantic Ocean and the North Sea.
- *Little Minch* .[301] 10 miles wide. Between the Outer Hebrides and the Island of Skye (United Kingdom), in the Atlantic Ocean.

---

[300]Obviously although this category of straits is denominated as "principal straits", not all of them are equally important. While some of these are the most important straits for international navigation, such as *Dover, Gibraltar, Bab-el-Mandeb, Malacca, Singapore* and *Hormuz*; however, some of the straits excepted from this regime, such as *Tiran, Foveaux* and *Messina*, are equally important.

[301]Attention should be drawn to the fact that we include the Little Minch in article 37 because it is located not between an island of a State and its mainland territory, but between two groups of islands, therefore, it does not correspond to the case in article 38.1, and it also joins two zones of

- *North Channel*. 16 miles wide. Between Scotland and Northern Ireland. It joins the Irish Sea and the Atlantic Ocean.
- *Dover Strait* or *Pas de Calais*. Its width varies between 17 and 26 miles. It is 30 miles long, and its depth varies from 20 to 37 m. It runs between the southeast coast of the United Kingdom and the north coast of France in the English Channel. It joins the Atlantic Ocean and the North Sea.
- *Strait of Gibraltar*.[302] This is 7.5 miles wide, 33 miles long and its depth varies from 10 to 100 fathoms. It is located between Spain and Morocco and joins the Atlantic Ocean and the Mediterranean Sea. Its direction is East-West.
- *Menorca Channel* or *Freu de Menorca*. 20 miles wide. Its depth varies from 24 to 80 fathoms. It separates the Islands of Mallorca and Menorca, in the Balearic Archipelago in Spain, in the Mediterranean Sea.
- *Strait of Bonifacio* or *Bouche of Bonifacio*. 6 miles wide and approximately 22 miles long. Its depth varies from 27 to 80 fathoms. It separates the Island of Corsica (France) to the north from the Island of Sardinia (Italy) to the south. It joins the Tyrrhenian Sea and the Mediterranean Sea.
- *Kerch Strait or Kerchenskly Proliv*.[303] This is approximately one mile wide, 25 miles long and 18 m deep. It is situated between the southwest coast of Russia to the east and the Ukrainian Peninsula of Crimea to the west. It joins the Black Sea and the Sea of Azov.
- *Kithera Channel (Dhiékplous Kithiron)*. 4 miles wide and 10 miles long. Its depth varies from 39 to 300 fathoms. It lies between the Greek Islands of Kythera and Antikythera. It joins the Mediterranean Sea and the Aegean Sea.
- *Karpathos Strait or Scarpanto Strait*. 23 miles wide and with a maximum length of twenty-one and a half miles in its eastern part. Its depth varies from 13 to 700 fathoms. It lies between the Greek Islands of Karpathos and Rhodes (in the Dodecanese). It is the most easterly of the straits which join the Mediterranean Sea and the Aegean Sea, and goes in a North-South direction.
- *Bab-El-Mandeb Strait* ("Gate of Tears"). Generally it is 19 miles wide, its length is approximately 50 miles, and its depth varies from 100 fathoms to 2 and 6 fathoms in the area of the coastal reefs. It is located between Yemen to the

---

the high seas or exclusive economic zones (the Atlantic ocean and the North Sea), even though its waters are internal on applying article 35 a). This means that passage in transit governs. Nevertheless, this is not the opinion of the United kingdom which, in its law of 1987 on territorial sea, only recognises passage in transit for the Dover Strait and The Hole, but maintains the right of innocent passage for the rest of its straits, which includes the Little Minch and the North Channel, and does not justify why it maintains the previous regime and does not apply Part III of the UNCLOS, which curiously it is a co-author of. The text of the British Law of 1987 can be read in the *Law of the Sea Bulletin*, No. 10, 1987, pages 11 et seq.

[302] The name Gibraltar derives from Jabel Tarik, the Moor who invaded the area in 711 AD.

[303] We should remember that Ukraine and Russia made a joint declaration, on December 24, 2003, that the waters of this Strait are historic, which would exclude it from the scope of article 37 by application of article 35 a). If this claim were accepted by the other States, then, the passage through the Kerch Strait would be subject to the consent of the coastal States, as is set out in this Declaration.

north and Djibouti and Somalia to the south. It joins the Gulf of Aden and the Red Sea. It forms part of the route from the Mediterranean Sea to the Far East. Perim Island divides the strait into two channels. Dact-el-Mayun is 10 miles wide and 10 miles long and Alexander's Strait, three and a half miles wide and three miles long.

- *Strait of Hormuz.* 21 miles wide and 100 miles long. It lies between Iran to the north and Oman and the Arab Emirates to the south. It joins the Persian Gulf and the Gulf of Oman.
- The straits of the Russian archipelago Franz Josef Land. *British Channel, Austrian Strait* and *Markham Strait*, in the Barents Sea.
- *Kara Strait.* 13 miles wide. It lies between the islands of Novaya Zemlya and Vaygach (Russia). It joins the Barents Sea and the Kara Sea.
- *Matochkin Strait.* This is approximately 600 m wide at its narrowest part, and is 55 miles long. It divides Novaya Zemlya into two (Severny Island and Yuzhny Island). It joins the Barents Sea and the Kara Sea.
- *Shokal'skii Strait* this is ten and a half miles wide. It lies between the Russian Islands of Bolshevik and Revolution. It joins the Kara Sea and the Laptev Sea.
- *Bering Strait.* A maximum of 45 miles wide between both mainland coasts, but there are islands which divide the strait into channels. There is a distance of 19 miles between Big Diomede and Siberia, between Little Diomede (U.S.A.) and Big Diomede there is a distance of 2 miles, and between Little Diomede and Alaska there is a distance of 20 miles. Its depth varies from 30 to 50 m. It separates Siberia (Russia) to the west and Alaska (U.S.A.) to the east. It connects the Bering Sea and the Arctic Ocean.
- *Pervyv Kuri'skiy Proliv.* 7 miles wide. It separates the Russian Peninsula of Kamchatka and the Island of Ostrov Shuneshu in the Kuril Islands. It joins the Ojotsk Sea and the Pacific Ocean.
- *Proliv Friza or Etorofu Kaikyo.* 20 miles wide. Between the Kuril Islands of Iturup or Etorofu and Urup. It joins the Ojotsk Sea and the Pacific Ocean.
- *Proliv Yekateriny or Kunasuiri Suido.* 10 miles wide. Between the Kuril Islands of Kunashir and Iturup or Etorofu. It joins the Ojotsk Sea and the Pacific Ocean.
- *Shikotan Suido.* 12 miles wide. It lies between the Minor Kuril Islands (also called Habomai) of Shikotan and Taraku. It joins the Ojotsk Sea and the Pacific Ocean.
- *Taraku Suido.* 12 miles wide. Between the Minor Kuril Islands or Habomai Islands of Taraku and Shibotsu. It joins the Ojotsk Sea and the Pacific Ocean.
- *Nemuro Kaikyo or Notsuke.* 10 miles wide. Between the Kuril Island of Kunashir Yekateriny and Japan (Hokkaido). It joins the Ojotsk Sea and the Pacific Ocean.
- *Tanega-shima Kaikyo.* 10 miles wide. Between the Japanese Islands of Tanega-shima and Yaku-shima, in the Ryukyu. It joins the East China Sea and the Pacific Ocean.
- *Tokara Kaikyo* or *Colnett Strait.* 22 miles wide. Between the Japanese Islands of de Osumi Gunto and Tokara Gunto in Ryukyu Archipelago. It joins the East China Sea and the Pacific Ocean.

- *Lema or Lamma Channel.* 6 miles wide. Between Hong Kong and the Lema Islands (China), in the South China Sea.
- *Malacca and Singapore Straits.* These are two straits which are a continuation of each other, therefore, they are considered to be a single sea lane and constitute the longest strait in the world with a total length of 500 miles. They connect the South China Sea and the Andaman Sea.
  The *Malacca Strait* separates Indonesia to the west and Malaysia to the east, and has a minimum width of 20 miles.
  The *Singapore Strait* separates Malaysia to the north and Singapore to the south, its minimum width is 8 miles. Dues to the fact that the State of Singapore is made up of a multitude of islands (more than 60), there are several channels in the Singapore Strait. Among these channels, attention should be drawn to the *Johor Strait* owing to its importance for international navigation. This is the northern channel of the Singapore Strait. It lies between the southern region of Johor (Malaysia) to the north and the Island of Singapore (Singapore) to the south. Its width varies from one to three miles and it is approximately 40 miles long.
- *Macpherson Strait.* Between the Islands of South Andaman and Rutland, in the Archipelago of Andaman and Nicobar (India). It joins the Andaman Sea and the Indian Ocean.
- *Palk Strait.*[304] 3 miles wide, between India and Sri Lanka, in the Indian Ocean.
- *Torres Strait.* Its navigable channel is 20 miles. It is located between Australia and Papua New Guinea. It connects the Arafura Sea and the Coral Sea.
- *Bass Strait.* Approximately 50 islands divide the strait into a number of channels, none of which is wider than 24 miles. It is located between the south of the Victoria Peninsula to the north and the Island of Tasmania to the south. It joins the Indian Ocean and the Tasman Sea.
- *Banks Strait.* 8 miles wide. It separates Cape Barren Island from the Island of Tasmania (Australia). It connects the Indian Ocean and the Tasman Sea.
- *Cook Strait.* 12 miles wide. Its depth varies between 20 fathoms and a little over 200. It separates North Island from South Island, in New Zealand. It joins the Tasman Sea to the north with the Pacific Ocean to the south.
- *Apolima Strait.* 4 miles wide. Between the Islands of Savati and Upolu in Samoa, in the Pacific Ocean.
- *Agattu Strait.* 17 miles wide. In the Aleutian Islands (U.S.A.). It connects the Bering Sea and the Pacific Ocean.
- *Tanaga Passage.* 19 miles wide. In the Aleutian Islands (U.S.A.). It connects the Bering Sea and the Pacific Ocean.
- *Unimak Passage.* 10 miles wide. In the Aleutian Islands (U.S.A.). It connects the Bering Sea and Pacific Ocean.

[304] As we mentioned above, India and Sri Lanka have declared that Palk Bay and Palk Strait include historic waters. A declaration which has been protested, therefore, we cannot consider this to be effective.

- *Samaiga Passage*. 16 miles wide. In the Aleutian Islands (U.S.A.). It connects the Bering Sea and the Pacific Ocean.
- *Seguam Passage*. 13 miles wide. In the Aleutian Islands (U.S.A.). It connects the Bering Sea and the Pacific Ocean.
- *Adak Strait*. 7 miles wide. In the Aleutian Islands (U.S.A.). It connects the Bering Sea to the Pacific Ocean.
- *Kaulakhi Channel*. 15 miles wide. Between the islands of Nihau and Kawai, in Hawaii (U.S.A.), in the Pacific Ocean.
- *Kaiwi Channel*. 22 miles wide and approximately 15 miles long. It has a depth of up to 35 fathoms. It is located in the Hawaiian Islands (U.S.A.), and it separates Oahu from Molokai, in the Pacific Ocean.
- *Pailolo Channel*. 8 miles wide. Between the islands of Molokai and Maui, in Hawaii (U.S.A.), in the Pacific Ocean.
- *Alalakeiki Channel*. 6 miles wide. Between the islands of Maui and Kahoolawe, in Hawaii (U.S.A.), in the Pacific Ocean.
- *Kealaikahiki Channel*. 15 miles wide. Between the islands of Lanai y Kahoolawe, in Hawaii (U.S.A.), in the Pacific Ocean.
- *Auan Channel*. 8 miles wide. Between the islands of Lanai and Maui, in Hawaii (U.S.A.), in the Pacific Ocean.
- *Kalohi Channel*. 8 miles wide. Between the islands of Molokai and Lanai, in Hawaii (U.S.A.), in the Pacific Ocean.
- *Barrow Strait*.[305] 15 miles wide. Between the islands of Prince Leopold and Melville in the Canadian Arctic Archipelago. It connects Lancaster Sound and *Prince of Wales Strait*, which has a width of 6.5 miles and runs between Banks and Victoria Islands, and finally joins the Gulf of Amundsen.
- *Robeson Channel* and *Nares Strait*. The former has a minimum width of 10 miles and the latter has a minimum of 14 miles, they are continuations of each other. They run between Ellesmere Island, in the Arctic Archipelago of Canada, and Greenland (Denmark). They connect the Lincoln Sea in the Atlantic Ocean and the Arctic Ocean.
- *Old Bahama Channel*. 16 miles wide. It is located between the north of Cuba to the south and Cay Lobos (Bahamas) to the north, in the Caribbean Sea.
- *Turks Island Passage*. 13 miles wide. Between the Turks and Caicos Islands (United Kingdom), in the Atlantic Ocean.
- *Guadeloupe Passage*. There are several islands which divide the strait into two channels which are no wider than 24 miles. At some points the depth reaches 1,000 fathoms. It is located between Guadeloupe (France) and Montserrat (United Kingdom). It joins the Atlantic Ocean and the Caribbean Sea.
- *Dominica Passage*. 16 miles wide and approximately 36 miles long. It separates Marie Galante (Guadeloupe, France) from Dominica. There are several islands

[305]We must not forget the objection mentioned above concerning compliance or not of the functional component, taking into account its merely temporary use for international navigation.

which divide the strait into a number of passages. It joins the Atlantic Ocean and the Caribbean Sea.

- *Martinique Passage*. 22 miles wide. Between Dominica and Martinique (France). It joins the Atlantic Ocean and the Caribbean Sea.
- *St. Lucia Channel*. 17 miles wide and a length of 8–13 miles. It separates Martinique (France) from St. Lucia. It joins the Atlantic Ocean and the Caribbean Sea.
- *Saint Vicent Passage*. 23 miles wide. Its length varies from four and a half miles to ten miles, and its depth is between 100 and 1,000 fathoms. It is located between St. Lucia and Saint Vincent. It joins the Atlantic Ocean and the Caribbean Sea.
- *Serpent's Mouth.*[306] 8 miles wide. Its depth varies from 5 to 20 fathoms. It runs between the Island of Trinidad (Trinidad and Tobago) to the north and the north coast of Venezuela to the south. It is funnel shaped and connects the Atlantic Ocean and the Gulf of Paria.
- *Dragon's Mouth*. 6 miles wide. It has a depth of 135 fathoms. It is located between the Island of Trinidad (Trinidad and Tobago) to the west, and the Paria Peninsula (Venezuela) to the east. It is the continuation of *Serpent's Mouth*. There is an area where some islands divide the strait into four channels. Boca de Monos, Boca de Huevos, Boca de Navíos and Boca Grande. It connects the Gulf of Paria and the Caribbean Sea.
- *Aruba-Paraguana Passage*. 15 miles wide. Between Aruba (Dutch Antilles) and the Paraguana Peninsula (Venezuela), in the Caribbean Sea.

If we again take the factors which indicate the importance of a strait for international navigation, namely, the tonnage of the vessels which cross the strait, the number and flags of these vessels, the size of the cargo, and its use by warships, we find that this last category of straits referred to, that of article 37 – passage in

[306]It is important to draw attention to the fact that article VI of the *Treaty of April 18, 1990*, signed by Venezuela and Trinidad and Tobago, *on the delimitation of sea and submarine areas*, expressly provides that "innocent passage is applied in the straits in the Gulf of Paria" (See *Law of the Sea Bulletin*, No. 19, 1991, p. 25). A provision which affects the *Serpent's Mouth* and the *Dragon's Mouth*, and fails to comply with article 37 of the UNCLOS according to which passage in transit is in force in these straits. If article 311.2 of the UNCLOS is applied, this treaty would not apply. However, although Trinidad and Tobago has ratified the 1982 Convention (April 25, 1986), Venezuela is not a party to this Convention and, according to the principle of *pacta tertiis nec nocent nec prosunt*, it is not obliged by its provisions (except for the provisions which are of a consuetudinary nature).

This involves an important controversy, taking into account that the 1990 Treaty is valid and legitimate for Venezuela as it is in accord with the consuetudinary norm on innocent passage codified in article 16.4 of the 1958 Geneva Convention; however, this is not so for Trinidad and Tobago which is a party to the UNCLOS and is obliged by its provisions. Evidently, Trinidad and Tobago has a problem concerning the application of successive treaties regarding the same matter which are incompatible with each other; the application of one entails the non-compliance of the other, which gives rise to its international responsibility as set out in article 30.5 of the *1969 Vienna Convention on the Law of Treaties*.

transit – includes what we could consider the five most important straits for international navigation, those which have a determining importance from a strategic and commercial point of view: *Dover, Gibraltar, Bab-el-Mandeb, Hormuz*, and *Malacca-Singapore*.

The *Dover Strait* is one of the busiest international straits, with approximately 350 vessels of all types using its main lanes every day. Due to its dense traffic and the rough weather and strong sea currents, there is a high risk of collision and approximately 2,000 losses have occurred.

The density of traffic in the Strait of *Gibraltar* is approximately 200 vessels daily, especially oil tankers, which transport about 200 million tons of crude oil each year. Navigation may be complicated when visibility is reduced. From the strategic point of view it is the most important strait as it is the natural passage into the Mediterranean Sea, and, consequently, to the Near East, while it provides an air corridor for aircraft which wish to cross the area and not fly over the air space of the adjacent States.

The strategic importance of the *Bab-el-Mandeb Strait* began with the opening of the *Suez Canal*, and this increased when the width and depth of the Canal were increased; approximately 50 vessels pass through this strait each day and 10% of the oil for Western Europe also passes through this strait each year.

The *Strait of Hormuz* is one of the most vital channels in the world for trade as two thirds of the sea transport of crude oil passes through this strait; approximately 80 vessels daily.

The economic and strategic importance of the *Malacca-Singapore Strait* has a long history, and it was controlled successively by Indians, Portuguese, Dutch and British until the independence of Malaysia, Indonesia and Singapore. Their importance grew with the opening of the Suez Canal in 1869, and, since 1950, it is the main artery for the shipment of crude oil in tankers from the Gulf of Japan. Approximately 140 ships cross these straits every day.[307]

There is no doubt that these are to blame for the convulsion arising in the III Conference of the United Nations on the Law of the Sea.

[307]Cf. *The Times Atlas and Encyclopedia of the Sea* (1983), p. 154–157.

Chapter 5
International Straits and the Right of Innocent Passage with No Possibility of Suspension

The *United Nations Convention on the Law of the Sea* of December 10, 1982 has supposed a revolutionary change in the regime of straits used for international navigation, which was substantially altered. Although it certainly consolidated some aspects of the legal regime in force up to then, for example, the legal 'indefiniteness' of this maritime area or the application of the right of innocent passage, although this applied to only some straits, this did not prevent the introduction of important innovations which entailed a radical change of regime.

In fact, a "purely textual or a broader contextual" interpretation of the UNCLOS, as pointed out by J.N. Moore,[308] clearly shows how the regulations set out for international straits differ very little from the regulations for the high seas as a regime of *freedom of navigation* through the straits is stipulated for the vessels of all States as a right of overflight, a right to underwater transit, rights of transit not subject to the character of 'innocent' required for territorial waters, and freedom of navigation for the maritime archipelago corridors are expressly recognised. This is fully adapted and favours the global, strategic interests which are at the root of the naval objectives of the grand maritime nations, in the same direct relation which distances itself from the consecrated sovereign rights of the coastal States of straits and damages their security.

Despite the fact that the new system adopted by the 1982 Convention does not, in any way, entail that the international straits have become independent maritime areas or different from the territorial sea of the 'strait' States. Although, basically, this may not seem so, there are no doubts in this regard, as article 34 of the UNCLOS eliminates such doubts when it initiates Part III by emphasising the maintenance of the sovereignty of the coastal State over the waters which form the straits used for navigation, evidently, for objectives other than those of

[308]See Moore (1980, pp. 120–121).

A.G. López Martín, *International Straits*,
DOI 10.1007/978-3-642-12906-3_5, © Springer-Verlag Berlin Heidelberg 2010

passage,[309] which leads us to think of a hypothetical exploitation of the natural resources in the straits.

Evidently the basic rule of the legal regime of Part III is the distinction between innocent passage with no possibility of suspension and transit passage; two distinct regulations intended for the straits used for international navigation. The key to the application of one or other system of passage or other as regards the coastal States of the strait and the States of the flags of the vessels and aircraft which cross these depend on the category the strait belongs to in accordance with the complex catalogue offered by Part III of the UNCLOS as seen in the previous chapter. We have passed from a single legal regime with no distinction as regards straits to a plurality of legal regimes which differentiates between 'straits' and 'straits'. From unity to homogeneity, we have gone from variety to heterogeneity or, perhaps, to discrimination?

In this regard, the last section of Part III is made up of a single provision, article 45, which regulates innocent passage through the straits used for international navigation. This is done from the perspective of *exception*.

The right of innocent passage, the cornerstone of the regime of navigation on territorial seas including international straits, the supreme rule for these seaways, has become a particular rule in an exceptional regime. In fact, the new 'transit passage' now becomes the regime of passage par excellence of the international straits, in the general rule, to which an astounding 8 of the 12 articles in Part III are dedicated, while innocent passage has been relegated to a merely residual level. Evidence of this is the fact referred to in the final article of the last section of this part:

1. The regime of innocent passage, in accordance with Part II, section 3, shall apply in straits used for international navigation:

   a) excluded from the application of the regime of transit passage under the article 38, paragraph 1; or
   b) between a part of the high seas or an exclusive economic zone and the territorial sea of a foreign State.

2. There shall be no suspension of innocent passage through such straits.

Two categories of international straits enter within the scope of the application of innocent passage: those made up of an island of a coastal State and its mainland territory, and those located between an area of the high seas or an exclusive economic zone and a territorial sea.

---

[309]Article 34 expressly provides the following:

1. The regime of passage through straits used for international navigation established in this Part shall not in other respects affect the legal status of the waters forming such straits or the exercise by the States bordering the straits of their sovereignty or jurisdiction over such waters and their air space, bed and subsoil.
2. The sovereignty or jurisdiction of the States bordering the straits is exercised subject to this Part and the other rules of international law.

The inclusion of this provision arises from the proposal submitted by the United Kingdom in the second session of the Conference, held in 1974. The British representative justified the difference of regulation alleging that:

> the interests of the international community in unimpeded navigation were not as strong in straits between the high seas and the territorial sea of a foreign State as they were in straits linking two parts of the high seas.[310]

It is evident that the British proposal and, therefore, Part III recognises and consecrates this 'difference' between straits.

In any case, article 45 does not develop the legal regime of innocent passage, but acts in remission of what is stipulated in this regard in the third section of Part II. That is to say, it refers us to the specific regulation of the territorial sea, corroborating that the waters which form part of an international strait continue to be territorial sea. There is only one particularity affecting the international straits: there is no possibility of suspending this passage, thus the common denomination of 'innocent passage which cannot be suspended'. Undoubtedly, in this point, the UNCLOS followed article 16.4 of the 1958 Geneva Convention.

In this regard, we find that the legal regime concerning innocent passage[311] is explained in detail in articles 17–32, which constitute the third section of Part II of the UNCLOS, more or less similarly to what is stated in the second section of Part I of the 1958 Geneva Convention. There are no substantial differences between both Conventions; however, it is certain that in the 1982 Convention there appeared specific regulations which affected the right of innocent passage, which in the conventional system of 1958 appeared under general formulas. This occurs with the criteria which breach innocent passage (article 19), with the material which the coastal State can regulate through domestic laws and regulations (article 21), or as regards seaways and the provisions for the separation of traffic (article 22).

5.1 Conceptual Aspects of Innocent Passage

L. Lucchini and M. Voelckel point out that "l'expression même: droit de passage innocent, ou inoffensif, surprend; la innocence n'est pas une 'donnée immédiate' de la conscience internationale, fut-elle juridique".[312] It is clear that we are not facing an innate virtue, but acquired or learned conduct, to which the consequent concession of certain rights is subordinated.

---

[310]See Second Committee, 3rd meeting (1974), p. 32, *II Off. Rec.*, 102.

[311]The regime of innocent passage today constitutes a totally accepted norm in the International Community already codified in the 1958 Geneva Convention. We do not wish to analyse the historical evolution of this right. An analysis of this right can be seen, among others, in Ghosh (1980, pp. 216–235); Ngantcha (1990); O'Connell (1984, pp. 259–298); Przetacznik (1977, pp. 222–236).

[312]See Lucchini and Voelckel (1996, p. 206).

The regime passage is also surprising as this is not simply a capacity, but an authentic right, which can be challenged by third States, and is a limitation of the sovereignty over territorial seas. This is a right which must be exercised in accordance with what is set out in the Convention, as stated in article 17, which serves as an introduction to Section 3:

> Subject to this Convention, ships of all States, whether coastal or land-locked, enjoy the right of innocent passage through the territorial sea.

In this way, the right of innocent passage must be exercised in a certain way and in accordance with certain conditions, stipulated in the UNCLOS.

From a reading of article 17, some conclusions can be deduced. One of these is that we are faced with a right which does not discriminate between States, as it attributes to all the States, whether coastal or not; the combined use of the expressions "all States" and "coastal or land-locked" supposes a clear redundancy which was incorporated into the 1958 Geneva Convention due to the insistence of the States with no coasts.[313] In addition, this expression also implies that this is a right which applies to the vessels of all the States which might not even be parties to the UNCLOS, as it is clearly a general norm of International Law declared by the Convention.

However, the preposition "of" must not be understood as indication 'ownership' of the vessel, but its 'nationality', a question which must be determined in accordance with the provisions in article 91 of the Convention.

Furthermore, the term "ships" must be evaluated within the whole of the Third Section of Part II as, in this Part, different types of ships are differentiated by establishing rules applicable to all vessels (articles 17–26), some applicable to merchant vessels and to State trading ships (articles 27 and 28), and others applicable to warships and other State ships which are not assigned to trade (articles 29–32).

The final important conclusion which we can deduce is that there is a regime applicable only to *sea* navigation, not to air navigation as only sea vessels come under this regime.

As regards *air navigation*, articles 1 and 2 of the *Chicago Convention on International Civil Aviation* of December 7, 1944,[314] which can be considered to be the General Law in force, they enshrine the sovereignty of the State as regards its air space, that is to say, as concerns the airspace of its territory in the strict sense, its internal waters and its territorial waters.

In this same regard, the Secretary of the ICAO has stated that no right of innocent passage is recognised for aircraft over territorial waters; this overflight is subject to the Chicago Convention. As regards this point, the 1982 Convention "does not change the *status quo* in international air law in this respect".[315]

---

[313]Cf. Nandan and Rosenne (1993, p. 156).

[314]The Chicago Convention came into being on April 4, 1947, and when Montenegro joined on February 12, 2007, there were 190 States ascribed. The text of the Convention can be seen at http://www.icao.int/icaonet/arch/doc/7300.html.

[315]See Doc. LC/26-WP/5-1 (1987), in *Netherlands Institute for the Law of the Sea, Documentary Yearbook*, 1987, p. 250.

'Innocent passage' is defined with regard to two inseparably joined factors: one is material and refers to the concept of passage; the other is the innocence of this passage. Two terms, passage and innocent, which have extremely complex components, which "show that it is quite difficult to define the combined words in precise terms",[316] but it is necessary to address these.

5.1.1 The Notion of Passage

Passage is understood as navigating through the territorial sea, regardless of whether this is to cross these waters without penetrating its internal waters, to head towards these waters or to head to the high seas coming from these waters. That is to say, it involves vertical, lateral and horizontal passage.

This first parameter draws attention to the activity which the vessel carries in the waters of another State, such as continually navigating through its territorial seas, without using the sea for other purposes than those strictly referred to navigation. Thus, stopping and anchoring are admitted only insofar as they are normal incidents of navigation or are imposed by force majeure or external danger.

In this regard, attention should be drawn to article 18 of the UNCLOS: "*Meaning of passage*"

1. Passage means navigation through the territorial sea for the purpose of:

   a) traversing that sea without entering internal waters or calling at a roadstead or port facility outside internal waters; or
   b) proceeding to or from internal waters or a call at such roadstead or port facility.

2. Passage shall be continuous and expeditious. However, passage includes stopping and anchoring, but only in so far as the same are incidental to ordinary navigation or are rendered necessary by *force majeure* or distress or for the purpose of rendering assistance to persons, ships or aircraft in danger or distress.

We should point out, with S.N. Nandan and Sh. Rosenne,[317] that, although the text refers to 'passage' in terms of 'navigation through the territorial sea', the meaning of passage contained in this provision is not limited to this section, but can be extrapolated to other parts of Convention on condition that it entails movement through a determined zone of the sea. Thus, passage is classified as "innocent" as regards the territorial sea, as 'transit' in international straits, and as 'archipelagic sea routes' as regards certain area s in archipelagic waters; all these cases refer to article 18.

In the light of article 18, and more specifically in paragraph (b), it can be presupposed that there is a right of States to access ports. As concerns this point,

[316]Cf. Ngantcha (1990, p. 56).

[317]See Nandan and Rosenne (1993, pp. 160–161).

O'Connell[318] points out that the inclusion of passage 'to and from ports' in internal waters is interpreted in the sense that this shows the existence of a rule of free access to ports, and that the coastal State is not free to deny the rights of navigation whose finality is this access. This interpretation is grounded on the comparison of this provision with the preceding article 14.2 of the 1958 Geneva Convention, which did not refer to the bays or the port installations. The origin of this innovation lies in a proposal presented by Fiji in 1973 to the Seabed Committee, and again submitted in the 1974 Conference, according to which, for the purposes of the articles on innocent passage, the term 'port' includes any port or bay installation which is normally used for loading, unloading and the anchoring of vessels.[319] However, this interpretation seems to be too extensive and not sufficiently grounded in practice and by the provisions in other parts of the Convention.[320]

The second paragraph of article 18 again takes up the principles enshrined in Geneva and adds the finality of providing assistance. However, it does include an interesting innovation, which is the introductory phrase, "passage shall be continuous and expeditious". This statement was initially proposed in 1974 by Malaysia, Morocco, Oman and Yemen, although this was not incorporated into the texts of the negotiation until 1976.[321]

As pointed out by T. Treves,[322] the inclusion of this statement seems to be intended to prevent interpretations in bad faith as regards the definition of the first paragraph by the States which exercise the right of passage. In addition, it allows the coastal State the possibility to question the exercise of this right in certain cases. Undoubtedly, this is a provision which is channelled towards the interests of the coastal States rather than the States which exercise the right of passage, although it is not a true innovation.

Finally, we should point out that the strategy which commenced at the 1958 Convention and continued at the 1982 Convention, with regard to providing a definition of 'passage' separated from the definition of its 'innocence', is ultimately intended to grant the coastal state "unquestionable powers to expel ships which are not in passage";[323] for example, when they stop or anchor and this is not a normal incident in navigation, nor is there force majeure or extreme danger.

---

[318]See O'Connell (1984, p. 269).

[319]Cf. Doc. A/AC.138/SC.II/L.42 and Doc. A/CONF.62/C.2/L.19, respectively.

[320]We must state that ports are part of the internal waters of States, consequently, these are subject to the territorial sovereignty of the coastal State, as stated by the ICJ in *Military and Paramilitary Activities in and against Nicaragua (Nicaragua v. United States of America) Case (merits)*, which also stressed that "it is also by virtue of its sovereignty that the coastal State may regulate access to its ports" (See ICJ *Reports*, 1986, pa. 213, p. 101).

[321]Cf. Doc. A/CONF.62/C.2/L.16, art. 3, p. 3, and Doc. A/CONF.62/WP.8/Rev.1, article 17, p. 2, respectively.

[322]See Treves (1985, pp. 754–755).

[323]See O'Connell (1984, pp. 269–270).

In this regard, the determination whether a vessel is or not in 'passage', in the sense expressed in article 18, we understand that this is the exclusive competence of the coastal State.

5.1.2 The Need for 'Innocence'

The UNCLOS again takes up the terms of the 1958 Geneva Convention when it mentions the "innocent" ("inoffensif" in French, "inocente" in Spanish) nature which passage must have in order to exist as a right for third part States. In the same way, it understands that "passage is innocent when it is not harmful to peace, good order or the security of the coastal State" (article 19.1 UNCLOS).

Undoubtedly, the expression used is ambiguous, imprecise and subjective as it leaves the decision on whether passage is harmful for its peace, order or security to the discretion of the coastal State. Thus, in order to achieve objectivity and specificity as regards the determination of the acts which, in fact, constitute harm, the second paragraph of article 19 lists 12 activities which are considered to be harmful and, consequently, are not innocent, therefore, they cannot be carried out by a vessel during its passage. This list did not exist in the conventional regime of 1958.[324] The acts stated in this context are the following:

(a) Any threat or use of force against the sovereignty, territorial integrity or political independence of the coastal State, or in any other manner in violation of the principles of international law embodied in the Charter of the United Nations
(b) Any exercise or practice with weapons of any kind
(c) Any act aimed at collecting information to the prejudice of the defence or security of the coastal State
(d) Any act of propaganda aimed at affecting the defence or security of the coastal State
(e) The launching, landing or taking on board of any aircraft
(f) The launching, landing or taking on board of any military device
(g) The loading or unloading of any commodity, currency or person contrary to the customs, fiscal, immigration or sanitary laws and regulations of the coastal State
(h) Any act of wilful and serious pollution contrary to this Convention
(i) Any fishing activities
(j) The carrying out of research or survey activities
(k) Any act aimed at interfering with any systems of communication or any other facilities or installations of the coastal State
(l) Any other activity not having a direct bearing on passage

If we examine the content and specific objective of the activities mentioned, as pointed out by T. Treves,[325] these refer mainly to the security requirement of the

[324]The only harmful act expressly referred to by the 1958 Geneva Convention is the passage of fishing vessels which fail to comply with the legislation of the coastal State (article 14.5).
[325]See Treves (1985, pp. 757–758).

coastal State which is stated in the general definition of the 'innocent' nature of passage. This is present in the first six sections; including, the first of these repeats what is expressed in article 301 UNCLOS,[326] which is directly inspired in article 2.4 of the *United Nations Charter* which includes the structural principle of the prohibition of the use or threat of force. The concern for security also appears to have a fundamental influence on sub-section k.

The other points on the list (g, h, i, j), however, refer to the sovereign activities of the coastal State, which are considered to have particular importance; including, some activities which are exercised exclusively by the coastal State beyond the territorial sea, as occurs with customs, taxation, health and immigration legislation, which applies in the adjacent zone (article 33 UNCLOS).

Although the catalogue of forbidden activities is rather extensive and detailed, we cannot consider that it is exhaustive or all inclusive due to the residual clause entailed by the final sub-paragraph, on referring to 'any other activity which is not directly related to passage'. The innovative intention of the Convention is clearly compromised with this final reference which, although it does not grant the use of absolute discretion to the coastal State, it must be interpreted and applied in consonance with the other points in the list in order to avoid the possibility of abuse of law, which is conventionally sanctioned in article 300 of the UNCLOS.

Furthermore, if we examine the activities listed, we can see that there is a mixture of definitive prohibitions as regards certain types of activities, such as the use of weapons, the launching or landing of any aircraft or military device, any act involving intentional and serious pollution or any fishing activity, together with others which are illegal only if the acts are carried out with a *specific intention* – as is the case of any propaganda act or the obtaining of information *which might harm* the defence or security of the coastal State, or any activity *aimed at* interfering in its facilities, installations or communication systems. The inclusion of a certain "intentionality" in the forbidden acts is precisely what gives rise to the subjective aspects which affect the classification as 'innocence', which this detailed list was intended to eliminate and these have not disappeared completely.

However, those which seem to be bluntly prohibited, or at least some of these, are bereft of this when article 19.2 is read in conjunction with article 21, as article 21 permits the coastal State to adopt laws and regulations as regards a number of activities, some of which correspond to acts forbidden under article 19.2; as occurs with fishing, maritime research or hydrographical surveys. Thus, these activities must be carried out with consent of the coastal State and in conformity with its laws

[326]Article 301 sets out:

Peaceful uses of the seas: In exercising their rights and performing their duties under this Convention, States-Parties shall refrain from any threat or use of force against the territorial integrity or political independence of any State, or in any other manner inconsistent with the principles of international law embodied in the Charter of the United Nations.

and regulations. When this happens, the passage cannot be classified as 'non-innocent'.

We also believe that it is important to draw attention to the fact that, as H. Caminos[327] points out, when placing emphasis on the work of the vessel, the Convention requires that the innocence its passage be judged only by its conduct and not by the purpose or destination of its voyage. Only when the foreign vessel is engaged in one of the activities listed in article 19.2, its passage can be considered as 'non-innocent', and this permits the coastal State to take the measures required to impede this passage as provided for in article 25.

As regards the classification of 'passage', the classification of 'non-innocence' of such passage is also the exclusive competence of the coastal State.

5.2 The 'Particular' Situation of the Passage of Certain Types of Vessels

As we have already stated, although it is true that article 17 attributes the right of innocent passage to the vessels of all the States with no distinction between the types of vessels, and that sub-section A of the third section of part II has the heading "Rules applicable to all ships", there are also certain particular categories of vessels which, in practice, involve certain reticence and discrepancy among the States as regards the automatic extension to these of the right of innocent passage. This is the case of the vessels which transport hazardous goods or those which are powered by nuclear energy, and warships. Added to these is the particular situation of submarines and submergible vehicles.

5.2.1 Exclusion of Submerged Navigation

One particular type of ship is the one which can sail while submerged, the submarine. In this case, innocent passage has an additional requisite which is included in article 20 of the UNCLOS:

> In the territorial sea, submarines and other underwater vehicles are required to navigate on the surface and to show their flag.

The inclusion of this norm implies that the right of innocent refers to surface navigation, and excludes navigation while submerged. This is a structural component of the notion of innocent passage.

---

[327]See Caminos (1987, pp. 141–142).

We must point out that this is strengthened by the provisions in article 24.1, which prohibits the coastal State from *de facto* or *de iure* discriminating against the vessels of any State or against those which transport goods to or from any State.

Undoubtedly we are dealing with a rule which is generally recognised in the Law of the Sea and was stated in the 1930 Hague Codification Conference, and as declared in article 14.6 of the 1958 Convention. However, this right only is referred to submarines while the corresponding article 20 of the 1982 Convention adds the reference to "other underwater vehicles", although it does not specify which vehicles are being referred to. Evidently, this inclusion is a response to technological and scientific progress in this are over the last few decades.

As concerns this particular situation involving these types of vessels, we must draw attention to the fact that, in the comment made by the International Law Commission on article 15 of its 1956 Draft Articles, it is stated that the provision was equally applicable to military and commercial submarines "if these ships are ever re-introduced".[328]

In addition, we believe that it is also important to stress the initial reference of article 20, "In the territorial sea". Bearing in mind that we are in Part II, which deals with the territorial sea, this reference seems to be redundant. However, this repetition is justified by the need to clarify that this provision is only applied to the zones of territorial sea which do not have a particular regime as is the precise case of the straits used for international navigation regulated by transit passage. In fact, the regulation of the exclusion of navigation while submerged is only related to innocent passage, as regards territorial seas in general and the international straits of article 45 in particular, but not as regards transit passage where this navigation is allowed (article 39.6 UNCLOS).

5.2.2 Vessels with Special Characteristics

The right of innocent passage of vessels which transport hazardous goods, tankers, as well as those powered by nuclear energy, have given rise to reticence in some States which understand that the harmless nature of the passage, inherent to the recognition of the right, cannot be appreciated if we forget the risk of contamination of the marine environment these vessels involve. The protection of the coast of the coastal State and, more specifically, its environment, have inspired the idea of a "modernisation of the notion of inoffensive passage",[329] defended tooth and nail by Canada, which would entail the recognition of special rights of the coastal State concerning the innocent passage of these vessels.

Results of this pressure are the two articles incorporated into the UNCLOS and which refer to these vessels with their special characteristics, articles 22.2 and 23.

[328]See ILC *Yearbook*, 1956, II, p. 270.

[329]Cf. Lucchini and Voelckel (1996, p. 260).

Article 22 sets out:

2. In particular, tankers, nuclear-powered ships and ships carrying nuclear or other inherently dangerous or noxious substances or materials may be required to confine their passage to such sea lanes.

Article 23 states: "*Foreign nuclear-powered ships and ships carrying nuclear or other inherently dangerous or noxious substances*".

Foreign nuclear-powered ships and ships carrying nuclear or other inherently dangerous or noxious substances shall, when exercising the right of innocent passage through the territorial sea, carry documents and observe special precautionary measures established for such ships by inte-rnational agreements.

In the light of these provisions, can it be said that these vessels do not have the right of innocent passage? Evidently, the response is negative. Innocent passage does not recognise the type of cargo or the type of propulsion as an exception. Furthermore, these articles confirm that this type of vessels has the right of innocent passage. In fact, the expression in article 23, "when exercising the right of innocent passage", and the reference of article 22 to the capacity "to confine their passage to such sea lanes", corroborate that the right of innocent passage for vessels powered by nuclear energy or which transport hazardous substances exists. These articles regulate their passage.

However, these provisions and specifically article 23 have been interpreted in a number of ways, just as the exercise of innocent passage by this type of vessels has been the subject of much pressure in contemporary practice.[330]

In this regard, the most controversial aspect of article 23 refers to the exact meaning and scope of allusion to the 'international agreements'. Does it refer to the international agreements to which the States are parties? One interpretation in this regard would greatly limit the efficacy of this legislation as it would only allow those agreements involving the coastal States and the flag States, and all those treaties which both States are not parties to would be discarded. In the opinion of L. Lucchini and M. Voelckel, the most reasonable interpretation is to understand that reference is being made to the agreements "généralement acceptés".[331] This supposes the application of the treaties the States are parties to and the others they are not parties to, but which codification of the norms are generally accepted as law. One of these international agreements would be the *International Convention for the Safety of Life at Sea (SOLAS)* of 1974, in particular its chapter VIII concerning nuclear powered vessels, and chapter VII regarding the transport of hazardous goods.[332]

[330]As regards this specific aspect, see Treves (1985, pp. 120–124).

[331]See Lucchini and Voelckel (1996, p. 262).

[332]Many other international agreements have been adopted in this area within the framework of the IMO, including the *Security Code for Nuclear Merchant Ships*, the *Fundamental Safety Principles* of IMO/IAEA, the *International Safety Management* (ISM Code), the *International Ships and Port Facility Security Code* (ISPS Code), the *International Code for the Construction and Equipment of Ships Carrying Dangerous Chemicals in Bulk* (IBL Code), or the *International Code for the Safe Carriage of Packaged Irradiated Nuclear Fuel Plutonium and High-Level Radioactive Waste on Board Ships* (INF Code, 2001). These texts can be seen in http://www.imo.org.

As we pointed out above, contemporary practice is not fully in consonance with this. Thus, Egypt[333] interpreted article 23 as permitting the States which are not parties to the aforementioned international agreements to submit the passage of these types of vessels to previous authorisation. Likewise Oman and Yemen[334] stated the same on ratifying the UNCLOS. Malaysia also submitted a declaration, accompanying its ratification, in which sustained that nuclear powered vessels and those which transported nuclear material required previous authorisation in order to enter a territorial sea.[335]

The *internal legislation* in this regard is more controversial as several States demand the authorisation or previous notification in order to allow these types of vessels to pass through their territorial waters. Article 5 *of Federal Law No. 19* of 1993, concerning the delimitation of the maritime zones of the United Arab Emirates,[336] establishes that vessels transporting hazardous or harmful substances have the obligation to notify of their passage. Likewise, in its *Law 52/AN/78*, Djibouti has included that such vessels must provide previous notification,[337] as is also established by article article 8 of *Law No. 45* of 1977 of the Democratic Republic of Yemen[338] regarding nuclear driven vessels and those which transport nuclear or radioactive substances, and the Pakistani *Law on territorial Waters and Maritime Zones* of December 31, 1976.[339]

The legislation in other States is more restrictive since they demand previous authorisation so that these vessels might cross their territorial waters. This is the case of *Law No. 18 on Maritime Zones*, of August 25, 1999, of Samoa (article 14.2);[340] *Law on Marine Area* of the Islamic Republic of Iran of 1993 (article 9);[341] *Law No. XXIV* of 1978 of Malta; and *Law No. 6*/96 of the Maldives, concerning Marine Zones, June 27, 1996.[342] At first, France also subordinated the passage of nuclear vessels to the need for previous authorisation in order to enter interior waters, territorial waters and French ports, pursuant to a Decree of June 19, 1969; however, the reference to territorial waters was suppressed in 1982.

---

[333] Declaration made when depositing its ratification instrument, on August 26, 1983. The text can be seen at http://untreaty.un.org.

[334] Declarations made at the time of its ratification (Oman, August 17, 1989, and Yemen, July 21, 1987). The text can be seen at http://untreay.un.org.

[335] The ratification of Malaysia took place on October 14, 1996. The text of the declaration can be seen at http://untreay.un.org.

[336] Cf. *Law of the Sea Bulletin*, No. 25, 1994, p. 93 et seq.

[337] Cf. http://www.un.org/Depts/los/LEGISLATIONANDTREATIES/regionalist.htm to read the text of the 1978 Law.

[338] Cf. *Ibidem.*

[339] Cf. *Ibidem.*

[340] Cf. *Ibidem.*

[341] Cf. *Law of the Sea Bulletin*, No. 24, 1993, pp. 10 y ss.

[342] Cf. http://www.un.org/Depts/los/LEGISLATIONANDTREATIES/regionalist.htm.

Undoubtedly, although the most interesting legislation for the purposes of ours study is the legislation adopted by coastal States of straits included in article 45, that is to say, *straits governed by the right of innocent passage*. In this regard, we must cite **Estonia**, a coastal state of the *Irben Strait, and of the Hari and Suur Straits, and of the Entrance to the Gulf of Finland*, all in the scope of article 45. According to its internal legislation, nuclear driven vessels must request previous authorisation through diplomatic channels, no less than 30 days before passage in order to be able to sail through its territorial waters and, consequently, to cross these straits. This is set out in *Decision No. 62* on the navigation of vessels through the territorial sea and the interior waters of the Republic of Estonia, of March 10, 1993:

> 5. For the passage of nuclear-powered ships through the territorial sea of Estonia the operator of the nuclear-powered ship has to submit an application to the Government of the Estonian Republic, making use of diplomatic channels, not later than 30 days before the planned passage. The Government of the Estonian Republic communicates its decision not later than 14 days before the planned passage.[343]

In addition, we must take into account the Decree of May 8, 1985[344] adopted by **Italy** as regards the *Strait of Messina*, whereby indefinite obligatory piloting is established for vessels which transport hazardous substances and weigh over 6,000 tons, and vessels of this type weighing 50,000 tons is directly forbidden.

Other internal legislation,[345] however, does not subordinate the passage of these vessels and entry to interior waters to the need for authorisation or mere notification although compliance with some requirements and information is required and these go beyond the mere obligation to use the sea lanes designated by the coastal State and to verify the requirements set out in article 23 of the UNCLOS.

We must state that all these positions have been the subject of the corresponding protest by the United States.[346]

5.2.3 Innocent Passage of Warships

Although there is no doubt that the most controversial issue concerning the application of the right of innocent passage to certain categories of vessels refers to

---

[343]See *Ibidem*.

[344]Cf. *Gazzetta Ufficiale* No. 110, de 11 de mayo de 1985.

[345]As an example, we can cite the French Decree of March 24, 1978, the French Law of January 2, 1979, the German Decree of November 12, 1984, or the Canada's Oceans Act of December 18, 1996, among others (the texts can be found at http://www.un.org/Depts/los/LEGISLATIONAND TREATIES/regionlist.htm).

[346]Cf. *Limits in the Sea*, No. 36, 8th revision, May 25, 2000.

warships.[347] A substantial number of States are much more reticent to the right of innocent passage for these types of ships, as is a part of the doctrine. In this regard, this attitude takes into consideration that this right is the corollary of the freedom and the needs of trade, and not of the freedom of navigation; neither fishing vessels nor warships have any relationship with this freedom.[348] While the Grand Sea Powers are the most fervent defenders of the existence of this right.

As O'Connell pointed out, both the 1958 Geneva Convention and the 1982 Convention "are ambiguous on the question of the right of innocent passage of warships".[349]

It is well known that the ILC included an article in its 1956 Project which allowed the coastal State to submit the passage of warships through its territorial sea to the requirement involving previous notification or authorisation (article 24).[350] However, it is also known that this article did not obtain the two thirds majority required to be approved, therefore, the Geneva Convention is silent on this point. Only the literal interpretation of its article 14.1 when it refers to "ships of all States" makes it possible to extend this to warships.

In this regard, practice subsequent to the 1958 Geneva Convention does not dispel any doubts in this area as the States are aligned in the two extreme positions: that of the need for authorisation and that of the recognition of the right of innocent passage for these vessels. This situation continued throughout all of the III Conference, where there were also attempts to incorporate a provision which would allow the coastal State to demand previous authorisation or notification from these ships,[351] and all these came up against the wall of the Grand Sea Powers, therefore, there are no provisions in this regard in the UNCLOS either.

Although it is certain that neither of the conventional regimes has an article which refers expressly to the passage of warships in order to expressly permit this nor subject it to the need for authorisation or notification. It is also certain that the evident interpretational doubts arising from the 1958 Geneva Convention in this

---

[347]The concept of "warship" is provided by article 29 UNCLOS, in the following terms:

> For the purposes of this Convention, "warships" means a ship belonging to the armed forces of a State bearing the external marks distinguishing such ships of its nationality, under the command of an officer duly commissioned by the government of the State and whose name appears in the appropriate service list or its equivalent, and manned by a crew which is under regular armed forces discipline.

[348]Cf. Lucchini and Voelckel (1996, p. 249).

[349]See O'Connell (1984, p. 274); the work we refer to for an analysis of the historical evolution of the doctrinal discussion and of State practice in this regard (pp. 274–294).

[350]Cf. ILC *Yearbook*, 1956, II, p. 276.

[351]This is the case of the proposal presented by Malaysia, Morocco, Oman and Yemen (cf. Doc. A/CONF.62/C.2/L.16, article 15.3); of a document presented in 1978 by Argentina, Bangladesh, China, Ecuador, Madagascar, Pakistan, Peru, the Philippines and Yemen; as well as several amendments presented in the course of the Conference in 1982, which were intended to include a sub-section to article 21 (which includes the legislative power of the coastal State), in the sense that it would be allowed to demand previous notification or authorisation from warships.

regard do not seem to exist in the 1982 Convention or, at least, there is sufficient data to clear up these doubts.

In fact, several arguments sustain the interpretation that the articles of the UNCLOS stipulate the innocent passage of warships. On the one hand, there is the reference in article 17 to "ships of all States" without distinction; on the other hand, the very title of subsection A: "Rules applicable to all ships", where innocent passage is located, which does not differentiate between types of ships, but the following subsections (B and C) do. Moreover, certain activities listed in article 19 as 'non-innocent' can only be done by warships, which means that military navigation is their objective; this is the case of "exercise or practice with weapons" (b), "launching, landing or taking on board of any aircraft" (e), or "any military device" (f). It would be absurd to include these activities if the innocent passage of warships is not allowed as these are the types of ships which carry out these activities.

However, State practice continues to be unclear, nor is it unanimous, and the States have opposing attitudes.

When signing or depositing their UNCLOS ratification instruments,[352] several States declared that the passage of warships through territorial waters is subject either to previous notification or to the need for authorisation. This is the case of China, Croatia, Egypt, Montenegro and Serbia, which require previous notification. Iran, Oman and Yemen, on the other hand, sustain the need of previous authorisation for passage. Other States have laid claim to the right to take the measures required to safeguard their security interests, including the possibility to adopt laws and regulations concerning the passage of warships; this has been done by São Tomé and Príncipe, Cape Verde, Rumania, Sudan, Finland and Sweden. The opposite position has been sustained by other States, which have clearly stated that they favour innocent passage of warships with no need for consent or notification; this is the case of Italy, the U.S.A.,[353] the United Kingdom, France, the Netherlands and Germany.

Likewise, a substantial number of States maintain rules which impose the previous authorisation of the coastal State or the need to notify it in order to allow the passage of warships (just as we have seen in the cases of nuclear driven vessels and those which transport hazardous substances). The States which demand previous *notification* in their internal legislation are the following: Denmark (*Royal Order No. 73* and the *Ordinance* of April 16, 1999, which regulates the entry of foreign warships and military aircraft to Danish territory in peace time), Estonia (*Decision No. 62*, of March 10, 1993), Finland (*Decree 656/80,* of January 1, 1981), Guyana (*Law No. 10 on Maritime Borders*, of June 30, 1977), India (*Law No. 80*, 1976, on Maritime Zones), Indonesia (*Regulation No. 8*), Korea (*Presidential*

[352]The texts of the declarations can be read at http://untreay.un.org.

[353]In 1989 even the U.S.A. and the USSR formalised a joint declaration which was attached to the uniform interpretation of the rules of International Law which regulate innocent passage, in which both stated that all vessels, including warships have the right of innocent passage (cf. *ILM*., 1989, pp. 1444–1447).

Decree No. 9162, of September 20, 1978), Mauritius (the Territorial Sea Act of April 16, 1970), Montenegro and Serbia (Law of July 23, 1987).[354]

A greater number of States demand the concession of a previous *authorisation* so that warships might access their territorial waters: Albania (*Decree No. 5384*, of February 23, 1976), Algeria (*Decree No. 63-403*), Antigua and Barbuda (*Territorial Waters Act No. 18*, of August 17, 1982), Bangladesh (*Territorial Waters Act No.* 26, of April 13, 1974), Barbados (*Act No. 26* of 1977), Brazil (*Decree No.* 35, 844, which required *permission for more than three warships at the same time*), Bulgaria (Decree No. 514, modified by *Decree No. 90* of 1951), Burma (*Territorial Sea and Maritime Zones Act No. 3*, of April 9, 1977), Cambodia (*Decree of the Council of State* of July 31, 1982), China (*Law on the Security of Maritime Traffic* of 1983), Congo (*Order No. 049/77*), United Arab Emirates (Federal Act No. 19, of October 17, 1993), Grenada (Territorial Waters Act No. 17, of November 1, 1978), Iran (*Act on the Marine Areas* of May 2, 1993), Maldives (*Maritime Zones Act No. 6/96*), Malta (*Act No. XXIV* of 1978), Pakistan (*Territorial Waters and Maritime Zones Act* of December 31, 1976), Poland (*Order of the Ministry of Defence* of 1957), Rumania (*Decree No. 39* de 1956), Saint Vincent and the Grenadines (*Act No. 15* of 1983), Seychelles (*Act No. 2* of 1999), Somalia (*Law No. 37* of 1972), Sri Lanka (*Law No. 22* of 1977 and the *Presidential Proclamation* of 1976), Sudan (*Act No. 106 on Territorial Waters and the Continental Shelf*, of December 31, 1970), Sweden (Law No. 374 and Royal Decree No. 366), Syria (*Legislative Decree No. 304* de 1963 and *Law No. 37* of 1981), Vietnam (*State Decree No. 30/C* of 1980), and Yemen (*Presidential Resolution No. 17* and *Law No. 45* of 1977).[355]

In antagonism to these States, the position of the others States is one of the following: or the right of innocent passage for warships is expressly recognised in their legislation (Dominica, Spain, the United Kingdom, Russia and Ukraine), or this is based on the principle of reciprocity (Lithuania, Law of March 26, 1992), or they are simply silent as regards this matter.

However, evidently, for the purposes of this study, we are interested in verifying whether any 'particular points' arise as regards the recognition or not of innocent passage for warships when this occurs in *international straits*. One decision which inevitably springs to mind is that of the ICJ in the *Corfu Channel Case*. In its decision of April 9, 1949, the Court stated the existence of the right of innocent passage for warships,[356] as we have seen in the first chapter.

However, what is the position of the coastal States of the straits on this point? If we re-examine the contemporary practice we referred to above, we find that some

---

[354]The texts of the internal legislation mentioned can be found at http://www.un.org/Depts/los/LEGISLATIONANDTREATIES/regionalist.htm.

[355]Cf. *Ibidem*.

[356]Cf. ICJ *Reports*, 1949, pp. 28–29.

Coastal States of article 45 straits have clearly stated their support for the innocent passage of warships and they have even expressly included this in their internal legislation. This is the case of the **United States**[357] (*Etolin, Shelikof, Icy Straits, the Santa Barbara Channel, Juan de Fuca Strait and Head Harbour Passage*), **The United Kingdom** (*The Solent, the Pentland Firth*), **Italy** (*Straits of Messina Strait and Elba*), and **Russia** (*Kil'din, Proliv Litke, Proliv Nevel'skogo, Tartaria Straits*).

The position of the other States which demand previous notification from warships wishing to cross their territorial waters but exclude compliance with this prerequisite in order to cross the international straits where they are the coastal states. This is the case of **Denmark** and **Korea**. In the first case, this involves the *Great Belt*, the *Little Belt* and the *Oresund*,[358] which is a logical consequence of the special conventional long standing regime these straits are subject to, therefore this does not have any special significance. The position of Korea is more interesting as it is an article 38.1 coastal State, the *Cheju Strait*, which is regulated by the right of innocent passage according to the provisions in article 45. According to Presidential Decree No. 9162, of September 20, 1978,[359] the previous notification requirement for warships and government ships operated for non-commercial purposes is excluded in straits used for international navigation. Therefore, there is no doubt that warships and non-commercial State vessels have the right of innocent passage in the *Cheju Strait*. This may be a sample of a 'particular point' of international straits as regards navigation and innocent passage.

Nevertheless, this is not a generalised situation as there are other coastal States of straits regulated by innocent passage which do not automatically recognise this right for warships. This is the case of **Albania**, **Estonia, Finland**, **Sweden** and **Montenegro**. The first is a coastal State of the *Corfu Strait* and its Decree No. 5384 requires the previous authorisation of warships to enter its territorial waters despite the Court's decision of 1949. In the case of Estonia, we have already mentioned that it is a coastal State of the *Hari, Suur,* and *Irben Straits* and the *Entrance to the Gulf of Finland*; Decision No. 62, of March 10, 1993, requires previous notification for the passage of warships, through diplomatic channels. This same notification requirement is included in Law No. 463/56 and Decree 656/80 of Finland, which is a coastal State of the Entrance to the Gulf of *Finland*. Montenegro is a coastal

[357]The United States protested against the declarations made on ratifying or signing the UNCLOS, demanding previous notification or authorisation of warships in order to cross territorial seas, as well as the internal legislations which impose this prerequisite (cf. *Limits in the Sea*, No. 36, 8th revision, May 25, 2000).

[358]It should be pointed out that, although Royal Decree No. 73 of 1976 abruptly excluded the straits from the notification requirement, the Ordinance of 1999 imposes a limit to this exception as, according to article 3.2, the simultaneous passage of more than three warships of the same nationality through the Great Belt, Little Belt and the Oresund does require previous notification through diplomatic channels (cf. *Law of the Sea Bulletin*, No. 40, 1999, p. 56).

[359]Cf. http://www.un.org/Depts/los/LEGISLATIONANDTREATIES/regionalist.htm.

State of the *Nikola Strait*, and article 17 of the Law of July 23, 1987[360] establishes that warships which wish to exercise innocent passage through its territorial waters must previously notify of this, with a minimum of 24 hours' notice.

However, the position of Sweden is the most eloquent. As we have seen above, Sweden is one of the States which requires previous notification for warships, as is established in Law No. 374 and in Royal Decree No. 366, but these norms exclude this requirement in the *Oresund*. In this case, it might be mistakenly thought that Sweden has the same position as Korea, and that it considers navigation through international straits as a particular situation within the regime of innocent passage.

Nevertheless, this is not the conclusion which can be drawn if all the factors in play are taken into consideration. Sweden makes an exception of the *Oresund* because it is obliged to do so since this strait is regulated by the long standing convention, the Treaty of Copenhagen of 1857, which stipulates the free transit of all vessels by this and other straits, which also occurs with Denmark as we have seen above. This is not the only strait where Sweden is the coastal State. It is also the coastal State of the *Kalmarsund*, a strait regulated by innocent passage as it comes under article 38.1. Swedish legislation does not expressly except this strait from the notification requirement as it does with the Oresund, which needs an interpretation *sensu contrario*. Evidently, if it is not expressly excluded, this is because it is included, that is to say, this strait is regulated by the internal norm which requires previous notification to cross the territorial sea. Therefore, the warships which wish to pass through the *Kalmarsund* must previously notify the Swedish authorities.

Finally, we can conclude that, although the meaning of the Convention seems to be clear, it is not possible to reach a definitive conclusion regarding State practice.

5.3 Mutual Rights and Obligations Entailed by Innocent Passage

The coastal States of a strait used for international navigation have a number of *competences* and *powers* recognised by international legislation and these are generally derived from their territorial sovereignty over the coasts and waters of the strait. Nevertheless, sovereignty does not only generate rights, but involves a number of *obligations* concerning third party States which are stipulated in

[360]This law was adopted by the Socialist Federal Republic of Yugoslavia, but it continues in force in the territory of the new State of Montenegro. In addition, as we stated, when it presented its instrument of succession to the UNCLOS on October 23, 2006, Montenegro declared that "it considers that a coastal State may, by its laws and regulations, subject the passage of foreign warships to the requirement of previous notification to the respective coastal State and limit the number of ships simultaneously passing, on the basis of the international customary law and in compliance with the right of innocent passage (articles 17–32 of the Convention)". Cf. http://untreaty.un.org.

International Law. Evidently, the rights of the coastal States generate obligations for the States which use the straits and vice-versa, the rights of the users give rise to the corresponding obligations of the coastal States. This clearly shows that we are faced with the perennial conflict between two sovereignties, that of the coastal State and that of the State of the flag using the strait.

5.3.1 The Rights of the Coastal State as Regards Innocent Passage Through Its Territorial Sea, Including the International Straits

The Geneva Convention did not consider that it was necessary to expressly establish the legislative competences of the coastal State, in such a way that these were only guaranteed with reference to article 17, in the sense that the vessels which might exercise the right of innocent passage must comply with the laws and regulations adopted by the coastal State, in conformity with these articles and other regulations of General International Law, in particular, those concerning transport and navigation. According to article 16, the powers of application were restricted to the general power to take the measures required to prevent any passage which was not innocent in its territorial sea, as well as the right to adopt all the measures required to prevent any infringement of the conditions applicable to the admission of foreign vessels to interior waters, as regards the vessels sailing towards these waters.

There are no substantial differences regarding the rights of the coastal State in the conventional regime of 1982, however, the UNCLOS has considered that it was necessary to regulate *in extenso* the legislative or regulatory competences of the coastal State; this means that there are more detailed regulations while it maintains the situation concerning the rights of protection.[361]

5.3.1.1 The Regulatory Competences of the Coastal State

The coastal State exercises its sovereignty over its territorial sea and, consequently, the totality of the legislative and regulation competencies, except for respect for the right of innocent passage. This general appreciation is synthesised in a list of specific points concerning which the coastal State might adopt laws and regulations, which must be duly publicised as specified in article 21.1, throughout eight points.

This seems to be an exhaustive list in accordance with the literal interpretation of article 21 which establishes that the coastal State can dictate laws and regulations concerning innocent passage through the territorial sea, "in respect of all or any of

[361]As regards this question of the powers of the coastal State, cf. Yturriaga Barberán (1991a, pp. 248–265).

the following". If this is so, it would enormously limit the sovereignty of the coastal State over a part of its territory as all the matters on which a state can legislate are not included. It does not mention the competence to adopt customs, tax, immigration and sanitary laws and regulations, but it does stipulate the infringements to its laws and regulations in this area. How can a state adopt laws and regulations in order to prevent the infringements of a legislation whose competence is not expressly included in article 21? This would be impossible with a restrictive interpretation of an exhaustive list.

In any case, the list presented by article 21 refers, fundamentally, to the safety of navigation, the safeguarding of economic interests or the preservation of the fundamental interests of the coastal State, which can be considered to be a mirroring and a complement to the stipulations in article 19.2.

Specifically, the points are those listed below.

"(a) The safety of navigation and the regulation of maritime traffic"

This first section of article 21 states a fundamental concept which was incorporated into the Convention, such as "the primacy of global maritime navigation".[362]

In order to understand this first point in its totality, we must place this in relation to the stipulations in article 22, concerning the *"sea lanes and traffic separation schemes in the territorial sea"*. Taking into account the importance of this point for the coastal States of straits and for States using these straits, it will be the subject of a detailed analysis in the following section.

"(b) The protection of navigational aids and facilities and other facilities or installations"

This second paragraph is also related to the safety of navigation and the regulation of all the facilities and installations in its territorial sea by the coastal State; the reference to "other installations" may include artificial islands and structures whose purpose, among other things is the exploitation and exploration of natural resources. We understand that this reference is the adequate complement to subsection (k) of article 19.2, and, on the other hand, it must be placed in relationship with the following subsection.

"(c) the protection of cables and pipelines"

Evidently, the facilities, installations, cables and pipes may be seriously damaged by the vessels during their passage; therefore, it is necessary to protect these through the pertinent regulation of the coastal State. Here the finality is to protect the integrity of the systems and to preserve the marine environment (piping).

"(d) The conservation of the living resources of the sea", and *"(e) the prevention of infringement of the fisheries laws and regulations of the coastal State"*

The regulation of these matters is not completely satisfactory as J.A. Yturriaga Barberán[363] points out that, hand to say that the coastal State can dictate "laws and regulations" in order to prevent infringements of its fishing "laws and regulations"

---

[362]See Nandan and Rosenne (1993, p. 200).

[363]See Yturriaga Barberán (1991a, p. 225).

is a redundancy; however, this is too restrictive, taking into account that, while the coastal State exercises its sovereignty, it has the competence to adopt fishing laws and regulations and any activity related to fishing, not only to prevent infringements of its fishing laws and regulations. In addition, this power must be placed in relation to the obligation to maintain and conserve the living resources of the coastal State in the Exclusive Economic Zone and in the high seas, as in article 19.2(i).

"(f) The preservation of the environment of the coastal State and the prevention, reduction and control of pollution thereof"

This subsection must be related to the other provisions of the UNCLOS which are directly related to the contamination of the marine environment. This complements article 19.2, in its subsection (h), which any intentional and serious act involving contamination to be non-innocent. Moreover, it must be related to article 192, which establishes the general obligation of States to protect and preserve the marine environment, and to article 194, which sets out the obligation of the States to prevent, reduce and control the contamination of the marine environment as regards activities under its jurisdiction and in zones where it exercises sovereign rights. Moreover, the regulatory powers of the coastal State concerning the preservation of the marine environment are developed in articles 210 and 211.4, as regards marine contamination in general, and in articles 207.1 and 2, 208.1 and 3, and 212, for specific types of marine contamination.

In addition, the parallel provision related to transit passage, article 42.1(b), is much more restrictive as it is limited to the competence of the coastal State to put the international regulations applicable to unloading hydrocarbons, petroleum waste and other harmful substance into effect.

"(g) Marine scientific research and hydrographical surveys"

It is necessary to relate this competence with article 245, which grants the coastal State the exclusive right to regulate, authorise and carry out scientific marine research in its territorial sea. We must also take into account that Part XIII develops the right of the coastal State to regulate this matter in detail in the Exclusive Economic Zone and on the high seas. We understand that these provisions can be applied *mutatis mutandi* for scientific marine research in the territorial sea, including the straits.

"(h) The prevention of infringement of the customs, fiscal, immigration or sanitary laws and regulations of the coastal State"

We must repeat the criticism made regarding subsection (e), in the sense that the reference to the competence of the coastal State to dictate laws and regulations in order to dispose customs, fiscal, sanitary and immigration laws and regulations is unnecessary. It is also absurd to grant legislative competence in order to prevent infringements of legislation concerning a matter for which legislative has not been given. As J.A. Yturriaga Barberán points out, "to loop the loop".[364] However, this provision complements article 19.2(g), and must be related to article 33 which grants competences to the coastal State in this are in the adjacent zone.

[364]See Yturriaga Barberán (1991a, p. 256).

As a logical consequence to the recognition of the legislative competence of the coastal State, the last paragraph of article 21 establishes the *obligation of the foreign vessels* which exercise the right of innocent passage to respect all the laws and regulations dictated by the coastal State, as well as all the generally accepted international norms concerning the prevention of boarding vessels at sea. Obviously, this obligation is not limited to the legislative competences described in article 21, but must be limited to all the measures dictated by the coastal State as regards innocent passage through its territorial sea, including the international straits. Thus, the vessels of third party States must also respect the legislation dictated by the coastal State as concerns the sea lanes and the devices used to separate traffic, in accordance with the provisions in article 22; and nuclear driven vessels and those which transport hazardous substances must comply with the measures set out in article 23. They must, of course, adapt their navigation to the requirements of 'passage' and 'innocence' specifically stipulated in articles 18 and 19 of the Convention.

5.3.1.2 The Jurisdictional and Protection Competences

The power to dictate laws and regulations concerning innocent passage is not the only competence recognised for the coastal State, which also has the power to be able to act in the event that these are not complied with. The obligatory or punitive law which the coastal State might exercise as regards the vessels which navigate through its territorial sea, including the straits, is clearly linked to the regulatory competences.

As regards this point, the coastal State has the right to adopt the measures required to *impede* any passage which is not innocent in its territorial sea. As we have pointed out above, the classification as 'passage' and its 'innocence', is the exclusive competence of the coastal State and must be exercised following the parameters stated in articles 18 and 19 of the Convention. In consonance with the stipulations in article 301 of the UNCLOS, the measures to be adopted by the coastal State cannot entail the use of armed force except in the case of legitimate defence.

Furthermore, in the case of vessels which are sailing towards interior waters or to a port installation situated outside these waters, that is to say, harbours or ports in the sense of articles 11 and 12, their right to take the necessary measures to prevent non-compliance with the conditions these vessels are subject to in order to enter these waters or installations is also recognised. This second paragraph of article 25 must be related to the stipulations in article 211.3, which refers to the possibility that the coastal States might establish special requisites concerning the entry of foreign vessels to its interior waters and ports in order to prevent, reduce and control the pollution of the marine environment. In no case can these requirements damage the continued exercise of the right of innocent passage of a vessel.

Together with the protection rights regulated in article 25, we must refer to the *jurisdictional* powers which the coastal State can exercise. The implications of the right of innocent passage as regards the exercise by the coastal State of its jurisdiction *strictu sensu* over merchant vessels and the vessels of States assigned for commercial purposes are addressed by articles 27 and 28, which maintain the essence the meaning of the conventional regime of 1958, and refer respectively to the criminal and civil jurisdictions.

As concerns *criminal jurisdiction*, the regulation stipulated in the UNCLOS supposes that there is a proper balance between the jurisdiction of the coastal State and that of the flag State. In this regard, the first paragraph of article 27 only allows the coastal State to exercise its jurisdiction concerning offences committed on board a foreign vessel, in five cases: (a) that the offence has consequences in the coastal State; (b) that the offence is of such a nature that it might disturb the peace of the country or good order in the territorial sea; (c) that the captain of the vessel or a diplomatic agent or a consular official of the flag State has requested assistance from the local authorities; or (d) that such measures are necessary in order to repress the illicit traffic of narcotics or psychotropic substances. The measures to be adopted by the coastal State may involve the arrest of persons and the investigation of the events. This regulation is applicable as regards the vessels which are exercising the right of innocent passage in the lateral sense, that is to say, crossing the territorial sea without entering interior waters. In the case of a vessel which proceeds from interior waters (vertical passage), the second paragraph of article 27 permits the coastal State to take any measures authorised by its domestic law in order to make arrests and carry out investigations on board the vessel.

With regard to *civil jurisdiction*, article 28 includes the generally accepted rule that the coastal State should not stop or divert a foreign ship passing through the territorial sea for the purpose of exercising civil jurisdiction in relation to a person on board the ship (paragraph 1); nor may it levy execution against or arrest the ship for the purpose of any civil proceedings, save only in respect of obligations or liabilities assumed or incurred by the ship itself in the course or for the purpose of its voyage through the waters of the coastal State (paragraph 2). As occurs with criminal jurisdiction, the situation differs if the vessel proceeds from its interior waters as, in this case, the coastal State can detain the vessel or adopt precautionary measures. We must stress the difference of language used in paragraphs 1 and 2. The former states that the coastal State "should not stop...", while the latter chooses the expression "may not levy...", this means that, in the first case, the obligation is not absolute, but is a recommendation, while, in the second case, the obligation is clear and categorical.

Bearing in mind that the Convention only refers to criminal and civil jurisdiction, we should then ask what occurs in the other branches of domestic law. Does this silence mean that the coastal State is completely free to adopt any physical execution or conservation measures stipulated in its legislation? In principle, nothing seems to impede this[365] except for the general prohibition concerning the abuse of law enshrined in article 300.

[365]Cf. Riphagen (1984, p. 202).

Parte II ends with applicable norms, exclusively for *warships and other government ships operated for non-commercial purposes.* There is no doubt that these are particular types of ships due to their passage, as we have seen, and as regards the sanctioning competences the coastal State has concerning the infringements committed by these ships. In this case, if a warship or a government ship operated for non-commercial purposes fails to comply with the laws and regulations of the coastal State in relation to passage through the territorial sea, the coastal State has the competence to demand that it abandon its territorial sea and claim due international responsibility for the damage which this violation might have caused from the flag State. These specific measures are a logical consequence of the immunity which protects this special category of ships in international legislation.[366]

In the light of the above, one question remains in the air. What are the consequences when the violations of the laws and regulations adopted by the coastal State concerning innocent passage are committed by merchant ships or government ships operated for non commercial purposes? As occurred with the 1958, the 1982 Convention does not provide an answer to this question. This can be interpreted in the sense that the coastal State can adopt all the measures which its laws and regulations stipulate in this regard.

5.3.2 The Obligations Which Fall Upon the Coastal State in Relation to This Right of Passage

As a consequence of the sovereignty which the coastal State has regarding its territorial sea, and as we have verified, the coastal State has a number of rights which entail obligations for the vessels which wish to exercise the right of innocent passage in this territorial sea. However, this sovereignty also involves obligations. In fact, in relation to the foreign vessels which exercise the right of innocent passage in its territorial sea, including the straits, the coastal State is obliged to comply with specific obligations, which generate correlative rights for the users of the straits.

Basically, the obligations which affect all coastal States of international straits as regards the exercise of the right of innocent passage by foreign vessels can be summed up in two words: abstention and information.[367]

---

[366]This is stipulated in articles 30, 31 and 32 of the Convention, in subsection C, *Rules applicable to warships and other government ships operated for non-commercial purposes*, together with article 29 which establishes the definition of "warships".

[367]According to Ngantcha (1990, p. 172) the obligations in question do not entail positive measures, such as dredging its territorial sea, the construction of lighthouses or the elimination of remains from the navigation channels. Nevertheless, if the coastal State generates hazards for navigation, for example, the construction of artificial islands, then it is obliged to duly mark these and provide the means required to facilitate navigation.

5.3.2.1 The Duty to Abstain

Regarding the specific case of innocent passage through international straits, we must mention the first fundamental obligation to abstain included in article 45.2, "There shall be no suspension of innocent passage through such straits". A circumstance which leds to denominating this modality of passage through the straits as "non-suspendible" innocent passage. Undoubtedly, this norm demonstrates the particularity of navigation in international straits as what is a right of the coastal State in the general regime of innocent passage, as is the power of suspension (article 25.3), becomes an obligation in the particular regime of innocent passage through international straits: the obligation not to suspend. A norm already included in the 1958 Geneva Convention.

Despite the fact that this is a norm generally accepted as law, there have been examples of non-compliance, specifically by Italy in the *Strait of Messina*.[368] In 1980, within the work for the improvement of the high voltage line between Calabria and Sicily, it was necessary to replace six subterranean electricity cables running from one side of the strait to the other. By a Decree of the Ministry of the Merchant Navy on November 5, 1979, the Italian Government decided to restrict navigation in the strait to an area 1,000 m wide for a period of five days. On March 21, 1985, the Greek oil tanker *Patmos* and the Spanish oil tanker *Castillo de Monte Aragon* collided in the strait and there was a large crude oil spill. The Italian Ministry of the Merchant Navy then adopted a number of measures for the regulation of marine traffic in the strait, including two Decrees. The first was adopted on March 27, 1985, and was of a provisional nature. However, the second, issued on May 8, 1985, is of indefinite duration, and establishes the obligation of merchant vessels to navigate on the right of the device for the separation of traffic and to inform the maritime authorities of the cities of Messina and Reggio Calabria before entering the strait. In addition, piloting is obligatory for merchant vessels weighing more than 15,000 tons and for vessels weighing more than 6,000 tons if they transport oil or other substances which are harmful for the marine environment. Finally, navigation in the strait is forbidden for vessels which transport oil or other substances hazardous for the marine environment and weigh over 50,000 tons.

It is to find a legal justification for these measures adopted by Italy, fundamentally as regards the Decree of May 8, 1985. Firstly, because in no case does the Convention permit temporary suspension, and even less does it permit indefinite suspension. Secondly, because the limitation of passage through the sea lanes and devices for the separation of the traffic of vessels can only be established for reasons of safety (article 22), once these reasons end the limitation must cease; it cannot be indefinite. A similar limitation is only allowed for nuclear driven vessels or those which transport hazardous substances, but not for merchant ships. Finally, the

---

[368] Cf. Scovazzi (1995, pp. 149–150).

requirement of previous notification is incompatible with the right of innocent passage of merchant vessels.[369]

The second obligation which falls on the coastal State is parallel and complementary to the first although this second obligation is generally applicable to any kind of innocent passage in the territorial sea. This involves the obligation not to place obstacles to the exercise of innocent passage. This was the only obligation recognised in the 1930 Hague Codification Conference, is repeated in article 15 of the 1958 Geneva Convention, and is explicitly mentioned and extended in article 24.1 of the UNCLOS, which stipulates the following:

> The coastal State shall not hamper the innocent passage of foreign ships through the territorial sea except in accordance with this Convention. In particular, in the application of this Convention or of any laws or regulations adopted in conformity with this Convention, the coastal State shall not:
> a) impose requirements on foreign ships which have the practical effect of denying or impairing the right of innocent passage; or
> b) discriminate in form or in fact against the ships of any State or against ships carrying cargoes to, from or on behalf of any State.

This provision recognises that innocent passage can be hindered by the laws and regulations dictated by the coastal State concerning such passage, and obliges this State not to obstruct passage. As we have seen above, article 21 grants wide powers to the coastal State to dictate laws and regulations on innocent passage which can generate obstruction, denial or hindrance as regards the exercise of this passage. Without the protection of article 24, this possibility would be frequent. We must point out that this same obligation is repeated in article 211.4, which establishes that the laws and regulations adopted by the coastal State for the prevention, reduction and control of marine pollution in its territorial sea must not hinder the passage of innocent vessels, in accordance with Section Three of Part II.

The 1982 Convention, together with that of 1958, does not develop the measures which the State of the flag would have within its reach in order to impose its right of innocent passage in the event that this was denied by the coastal State. This hypothetical situation, which is more probable to happen in the case of warships due to the remarks mentioned above, would give rise to a controversy between both States, which would have to be extended to the obligation of a peaceful solution to the controversies deriving from the interpretation or application of the UNCLOS, established in article 279. However, it must not be underestimated that this moderation generates a certain risk that we will again find ourselves in the scenario of the threat or use of force, with actions and reactions, measures and countermeasures.[370]

[369]In fact, this Decree was protested by the United States (cf. *Limits in the Seas*, No. 112, pp. 70–71).

[370]However, we must remember the fact that the conflicts deriving from innocent passage have been fewer than was expected as shown by international practice.

In this regard, the decision of the ICJ of 1949 in the *Corfu Channel Case* provides us with information on the actions which the States of the flag might adopt in this context, especially if this involves warships. We should remember that, when the Albanian authorities rejected the right of passage of British warships through the North Channel of the Corfu Strait, the Court considered that the demonstration of force by the United Kingdom, which sent a small fleet to pass through the strait with the crews at combat stations, the explosion of mines and observation of coastal defences, was legitimate.[371]

The other obligation which arises from this law refers to the *de facto* or *de iure* prohibition to discriminate, therefore, national and foreign vessels must be treated equally. Likewise, the prohibition to discriminate between vessels which transport goods from or to any State is intended to protect against acts which openly discriminate as stated in article 24 (discrimination 'in the form'), and against acts which are not discriminatory in themselves, but which produce discriminatory effects ('de facto' discrimination).[372]

Together with the generic abstention duty formulated by article 24.1, we can find other specific expressions of the abstention obligation.

Thus, article 21.2 imposes an important limit to the regulatory power of the coastal State as regards innocent passage when it provides that such laws and regulations shall not apply to the design, construction, manning or equipment of foreign ships unless they are giving effect to generally accepted international rules or standards. This provision is a tribute to the protection of the integrity of world maritime navigation and must be placed in relation to article 211. As concerns the generally accepted international regulations and norms on this point, we must understand these to be the following, the *SOLAS Convention* of 1974, the *Convention Load* Line of 1966 (LL 66), the *Torremolinos P*rotocol of 1993 relating to the *Torremolinos International Convention for the Safety of Fishing Vessels* of 1977 (SFV 1977 and SFV PROT 1993), the *International Convention on Standards of Training, Certification and Watch-keeping for Seafarers* (STCW 1978), and for *Fishing Vessel Personnel* (STCW-F 1995), *MARPO*L 73/78, and the *International Convention on the Control of Harmful Anti-Fouling Systems* of 2001 (AFS 2001).[373]

In addition, article 26 forbids the coastal State to impose any levies on foreign vessels due to the mere fact that they pass through their territorial waters. This law admits an interpretation *sensu contrario* since, if levies are exclusively impeded by the fact that the vessel is in passage, it is understood that levies may be imposed for reasons other than passage. This idea is extended in the second paragraph, which permits levies on vessels as remuneration for determined services rendered to the vessels; such as piloting. Again there is no discrimination. What fails to remain

[371]See ICJ *Reports*, 1949, pp. 30–32.

[372]Cf. Nandan and Rosenne (1993, p. 226).

[373]Cf. The reference in the Division for Ocean Affairs and Law of the Sea (2004, p. 39). The texts of the conventions can be seen at http://www.imo.org.

clear in the text is whether general services, such as the maintenance of navigation aids, come under the expression "specific services rendered".

5.3.2.2 The Obligation Concerning Publicity and Information

The coastal State also has a number of obligations concerning the obligation 'to duly publicise' the measures it adopts together with other information.

Article 21.3 obliges the coastal State to duly publicise the laws and regulations it dictates concerning innocent passage with reference to the subjects stated in the first paragraph of this same article.

It is also obligatory to publicise the charts which show the sea lanes and the devices for separating sea traffic designated or prescribed in order to regulate its territorial sea, including the international straits.

It seems clear that the instrument which is used to provide due publicity is a choice which corresponds to the coastal State, which must be taken into account so that the information might be effective, and must reach other States, entities, authorities and persons which must be guided by this information.

Furthermore, the coastal State is obliged to inform of all the dangers which it is aware of and might threaten navigation in its territorial sea, including the straits. One obligation already stated by the ICJ in the *Corfu Channel Case*.[374] As regards the form of this information, article 24.2 no longer mentions 'due publicity' as in previous cases, but does mention "appropriate publicity". What is 'appropriate' will depend on the circumstances of each case, and is a decision which corresponds to the coastal State.

All these obligations concerning publicity and information are identical to those imposed as regards transit passage, as we shall see in the following chapter.

5.4 Regulation of the Safety of Navigation and Sea Traffic: Sea Lanes, Devices for the Separation of Sea Traffic and Other Safety Measures Adopted in Straits Regulated by Innocent Passage

As we pointed out the first regulatory competence which article 21 attributes to the coastal States as regards innocent passage refers to the possibility of adopting laws and regulations concerning the safety of navigation and the regulation of sea traffic. Undoubtedly, in order to analyse this competence in all its extension, we must place this provision in relation to article 22 which regulates the establishment of sea lanes and devices for the separation of traffic in the territorial sea. It is evident that the

[374]Cf. ICJ *Reports*, 1949, p. 22.

establishment of these lanes and devices are questions which clearly enter within the scope of the safety of navigation and sea traffic.[375]

In this regard, it is evident that the Convention recognises that the coastal State is competent to designate sea lanes and devices for the separation of traffic in its territorial sea, which obviously includes the international straits of article 45. According to article 22.3(a), in the exercise of this competence, the State can take into account the recommendations of the competent international organization which is the International Maritime Organization (IMO)[376] although the UNCLOS does not expressly mention this.

As regards this point, the most relevant provisions of the IMO concerning sea lanes contained in the SOLAS V/8 Regulation, modified in 1995, and in the Resolution of the Assembly A.572 (14), concerning the *General Provisions on Ships' Routeing*, amended by Resolution MSC.71 (69) of 1998. The devices for the separation of traffic are included in the *Convention on the International Regulations for Preventing Collisions at Sea* of 1972 (COLREG), of specific interest are Rules 1 (d) and 10.[377] However, this is a mere suggestion as it is the coastal State which adopts and designates these.

---

[375]Such measures have been adopted also in other straits that they are outside of the UNCLOS, for example in the Sound there is a traffic separation scheme (COLREG.2/Circ.59, 2007).

[376]The IMO is a specialised organisation of the United Nations, created in 1948 (its original name was the International Maritime Consultative Organization), and engaged in the drafting of measures concerning maritime safety. Its headquarters are in London, and currently (February 1, 2010) it has 169 Member States and three Associated Members. To date, it has adopted round about 50 agreements and protocols, as well as more than 800 codes and recommendations on marine safety, prevention of pollution and other related questions.

[377]Rule 1(d) sets out that, "Traffic separation schemes may be adopted by the Organization for the purposes of these Rules".

While Rule 10 establishes, *"Traffic separation schemes*

(a) This Rule applies to traffic separation schemes adopted by the Organization and does not relieve any vessel of her obligation under any other rule.

(b) A vessel using a traffic separation scheme shall:

(i) proceed in the appropriate traffic lane in the general direction of traffic flow for that lane; (ii) so far as practicable keep clear of a traffic separation line or separation zone; (iii) normally join or leave a traffic lane at the termination of the lane, but when joining or leaving from either side shall do so at as small an angle to the general direction of traffic flow as practicable.

(c) A vessel shall, so far as practicable, avoid crossing traffic lanes but if obliged to do so shall cross on a heading as nearly as practicable at right angles to the general direction of traffic flow.

(d) (i) A vessel shall not use an inshore traffic zone when she can safely use the appropriate traffic lane within the adjacent traffic separation scheme. However, vessels of less than 20 metres in length, sailing vessels and vessels engaged in fishing may use the inshore traffic zone.

(ii) Notwithstanding subparagraph (d) (i), a vessel may use an inshore traffic zone when en route to or from a port, offshore installation or structure, pilot station or any other place situated within the inshore traffic zone, or to avoid immediate danger.

(e) A vessel other than a crossing vessel or a vessel joining or leaving a lane shall not normally enter a separation zone or cross a separation line except:

(i) In cases of emergency to avoid immediate danger; (ii) To engage in fishing within a separation zone.

Other factors to be considered are the channels which are normally used for international navigation, the special characteristics of certain vessels and channels, and traffic density (article 22.3, sections b, c and d).

Here we find a fundamental difference concerning this same matter in the right of transit passage as, according to its equivalent in Part III, article 41, in the straits regulated by transit passage, the coastal States propose the sea lanes and the devices for the separation of traffic to the IMO, but it is the Organisation which adopts these. Thus, in the straits of the article 45 the sea lanes are freely adopted and designated by the coastal State while in the straits of the article 37 are designated by the IMO after an agreement with the coastal State.

Although the UNCLOS does not expressly include this, it is understood that the coastal State is equally competent to replace these lanes and devices at its discretion when the circumstances require this.

As stated in article 22.1, the coastal State also has competence to oblige foreign vessels to use these lanes and devices, but only when it is necessary for the safety of navigation; this means that interests apart from safety may not be invoked. As we pointed out above, the only exceptions are tankers, nuclear driven vessels and those which transport hazardous substances. The coastal State can demand that they limit their passage to these lanes, with no limitation (article 22.2). Although article 22 refers to "foreign ships" these must be respected by 'all' vessels included the vessels of the coastal State, this understanding is coherent with the logic and so that the lanes and devices might be effective.

Although the use of devices for the separation of traffic is not obligatory except in circumstances which involve safety and the captain of the vessel is responsible for deciding whether to follow the recommended route or not after evaluating the situation and circumstances, it will be understood that the vessels which navigate in a direction contrary to the one recommended, within the traffic guidelines fixed by the devices, are exposed to irrational risks due to the possibility that they might meet with a large number of vessels navigating in the opposite or almost opposite direction to them. Therefore, the vessels which do not wish to follow the

---

(f) A vessel navigating in areas near the terminations of traffic separation schemes shall do so with particular caution.

(g) A vessel shall so far as practicable avoid anchoring in a traffic separation scheme or in areas near its terminations

(h) A vessel not using a traffic separation scheme shall avoid it by as wide a margin as is practicable.

(i) A vessel engaged in fishing shall not impede the passage of any vessel following a traffic lane.

(j) A vessel of less than 20 metres in length or a sailing vessel shall not impede the safe passage of a power-powered vessel following a traffic lane.

(k) A vessel restricted in her ability to manoeuvre when engaged in an operation for the maintenance of safety of navigation in a traffic separation scheme is exempted from complying with this Rule to the extent necessary to carry out the operation.

(l) A vessel restricted in her ability to manoeuvre when engaged in an operation for the laying, servicing or picking up of a submarine cable, within a traffic separation scheme, is exempted from complying with this Rule to the extent necessary to carry out the operation".

recommended routes are advised to remain sufficiently separated from the exterior limits of the devices established and not to disturb the organised traffic.

The coastal States have not exercised this power to adopt separation devices or another safety system for navigation in all the international straits regulated by innocent passage. There are only **nine straits** where the coastal State or the IMO have exercised this power.

5.4.1 Pentland Firth

The IMO adopted certain measures as regards navigation. It had laid down that loaded oil tankers sailing towards or from Flotta and Scapa Flow must not sail through the Firth against the current, nor in adverse meteorological conditions or when there is reduced visibility (SN/Circ.159). Thus, the vessels which transport oil or other hazardous substances and which weigh more than 5,000 tons must avoid the area delimited for the geographical coordinates:

(1) 58°46′.73 N, 3°17′.69 W (Tor Ness)
(2) 58°55′.00 N, 3°50′.00 W
(3) 59°17′.00 N, 3°50′.00 W
(4) 59°28′.00 N, 3°15′.00 W
(5) 59°28′.00 N, 2°19′.00 W
(6) 59°24′.00 N, 2°09′.00 W
(7) 59°05′.00 N, 2°09′.00 W
(8) 58°50′.00 N, 2°35′.00 W
(9) 58°44′.15 N, 2°54′.90 W (Old Head)

Then around the coast of South Ronaldsay and Mainland to:

(10) 58°57′.84 N, 3°21′.11 W
(11) 58°55′.97 N, 3°21′.11 W

Thence along the coast of Hoy to (1).

It also recommends other routeing measures such as the use of North Hinder to German Bight route and vice versa for the following classes of ships:

- Tankers of 10,000 tons gross tonnage and upwards, carrying oils as defined under Annexe I to International Convention for the Prevention of Pollution from Ships, 1973, as modified by the Protocol of 1978 relating thereto (MARPOL 73/78)
- Tankers of 5,000 tons gross tonnage and upwards, carrying noxious liquid substances in bulk as assessed as categories A or B of Annexe II of MARPOL 73/78
- Tankers of 10,000 tons gross tonnage and upwards, carrying noxious liquid substances in bulk as assessed as categories C or D of Annexe II of MARPOL 73/78
- Ships of 10,000 tons gross tonnage and upwards, carrying liquefied gases in bulk

These ships are recommended to use the route are referred to resolution A.486 (XII), adopted on 19 November 1981, concerning the "Recommendation on the use of adequately qualified Deep Sea Pilots in the North Sea, English Channel and Skagerrak" (SN/Circ.167).

5.4.2 North Minch

As a safety measure, the IMO forbids oil tankers weighing more than 10,000 tons gross tonnage to pass through the *Minches* except due to stress of weather or any other case of "force majeure" (SN/Circ.159).[378]

5.4.3 Entrance to the Gulf of Finland

Estonia unilaterally adopted several measures for the safety in the approaches to Tallin in 2005:

- Light (*Estonian Notice 12/133/05*): 59°28′.13 N, 24°49′.24 E
- Obstructions (*Estonian Chart 610*):
  Depth 6,4. 59°31′.30 N, 24°35′.20 E
  Depth 19,9. (a) 59°29′.62 N, 24°39′.15 E; (b) 59°29′.90 N, 24°45′.95 E
- Pilot boarding places (*Estonian Notice 12/129/05 y 12/130/05*):
  59°29′.0 N, 27°42′.0 E
  59°39′.10 N, 24°45′.00 E (Muuga)
  59°36′.80 N, 24°.37′.40 E (Muuga)

The IMO also established two devices for the separation of traffic in the Gulf of Finland (COLREG.2/Circ.51):

(a) One near Gogland Island which consists of two parts,

Part I consists of two traffic lanes separated by a zone with a centre line connecting the following geographical positions:

(1) 59°59′.00 N, 026°57′.40 E
(2) 59°58′.52 N, 027°03′.10 E
(3) 59°59′.47 N, 027°06′.30 E

The traffic separation zone is 0.5 mile wide. The traffic lanes on the both sides of the traffic separation zone are 1 mile wide. The direction of navigation will be 99°–279° and 59°.3–239°.3.

[378] The measure affects the two *Minches*, *North* and *Little*, although innocent passage regulates the former (article 38.1) and transit passage regulates the latter (article 37).

Part II consists of two traffic lanes separated by a line connecting the following geographical positions:

(1) 59°59′.47 N, 027°06′.30 E
(2) 60°07′.55 N, 027°32′.80 E

The traffic lanes on the both sides of the traffic separation zone are 1.25 miles wide. The direction of navigation will be 59°.3–239°.3.

(b) Another, near Sommers Island, which consists of four parts.

Part I consists of a roundabout around the separation zone 0.5 mile in diameter centred on the geographical position 60°11′.50 N, 027°46′.20 E. The roundabout lane is 1 mile wide.

Part II consists of two traffic lanes separated by a line connecting the following geographical positions:

(1) 60°07′.55 N, 027°32′.80 E
(2) 60°10′.77 N, 027°43′.62 E

The traffic separation zone is 0.5 mile wide. The traffic lanes on the both sides of the traffic separation zone are 1 mile wide. The direction of navigation will be 59°.3–239°.3.

Part III consists of two traffic lanes separated by a line connecting the following geographical positions:

(1) 60°11′.15 N, 027°49′.05 E
(2) 60°07′.70 N, 028°16′.10 E

The traffic lanes on the both sides of the traffic separation zone are 1 mile wide. The direction of navigation will be 104°.3–284°.3.

Part IV consists of two traffic lanes separated by a line connecting the following geographical positions:

(1) 60°12′.70 N, 027°47′.90 E
(2) 60°24′.54 N, 028°05′.05 E

The traffic lanes on the both sides of the traffic separation zone are 0.5 mile wide. The direction of navigation will be 35°.7–215°.7.

Also establishing of deep water route inside the border of the traffic separation scheme from the Gogland Island to the Rodsher Island. The route lane is 1,000 m wide with established direction of traffic flow and is intended for the passage of ships with a draught up to 15 m:

1. Route: 60°01′.55 N, 027°11′.20 E; 59°59′.12 N, 027°03′.05 E. Direction: 239°.3. Distance: 4.8 miles. Lane width: 5.4 cables.
2. Route: 59°59′.12 N, 027°03′.05 E; 59°59′.90 N, 026°53′.57 E. Direction: 279°. Distance: 4.8 miles. Lane width: 5.4 cables.
3. Route: 59°59′.90 N, 026°53′.57 E; 60°03′.25 N, 026°40′.00 E. Direction: 296°.5. Distance: 7.6 miles. Lane width: 5.4 cables.

4. Route: 60°03′.25 N, 026°40′.00 E; 60°02′.06 N, 026°30′.30 E. Direction: 255°.5. Distance: 5 miles. Lane width: 5.4 cables.

COLREG.2/Circ.52 inserts the following amendments to the existing traffic separation schemes in the Gulf of Finland:

(a) Traffic separation scheme "Off Porkkala Lighthouse".

A separation zone, 1 mile wide, is centred upon the following geographical positions:

(1) 59°43′.70 N, 024°14′.00 E
(2) 59°44′.90 N, 024°21′.40 E
(3) 59°45′.90 N, 024°31′.00 E

A traffic lane, 1.5 miles wide, is established on each side of the separation zone.

A precautionary area is established upon the following geographical positions:

(1) 59°43′.95 N, 024°31′.80 E
(2) 59°46′.15 N, 024°53′.50 E
(3) 59°50′.05 N, 024°51′.90 E

A separation zone, 1 mile wide, is centred upon the following geographical positions:

(1) 59°48′.10 N, 024°52′.70 E
(2) 59°48′.80 N, 025°00′.00 E

A traffic lane, 1.5 miles wide, is established on each side of the separation zone.

(b) Traffic separation scheme "Off Hankoniemi Peninsula"

A separation zone, 2 miles wide, is centred upon the following geographical positions:

(1) 59°24′.50 N, 022°25′.00 E
(2) 59°28′.00 N, 022°34′.00 E
(3) 59°30′.00 N, 022°45′.00 E

A traffic lane, 4 miles wide, is established on each side of the separation zone.

5.4.4 Strait of Elba or Piombino

Italy has placed three lights in the Canale di Piombino in 2005 (*Italian Notice 22.6/05*) upon the following geographical positions:

(1) 42°51′.9 N, 10°28′.6 E
(2) 42°52′.0 N, 10°28′.9 E
(3) 42°51′.9 N, 10°28′.5 E

5.4.5 *Strait of Tiran*

The IMO adopted a device for the separation of traffic at the entrance to the Gulf of Aqaba that entered into force October 1, 1993 (COLREG.2/Circ.39/Add.1). Moreover, in order to safeguard the wrecked ecosystem, the environment and the economy of the zone, it laid down that all vessels carrying dangerous or toxic cargoes, or any other vessel exceeding 500 gross tons should avoid the area bounded by lines connecting the following geographical positions (SN/Circ.173):

(1) 28°46′.0 N, 34°37′.5 E
(2) 28°46′.0 N, 34°40′.0 E
(3) 28°24′.0 N, 34°31′.0 E
(4) 28°18′.0 N, 34°26′.0 E
(5) 28°11′.0 N, 34°29′.0 E
(6) 28°06′.0 N, 34°28′.0 E
(7) 28°01′.5 N, 34°26′.5 E

5.4.6 *Strait of Georgia*

The device for the separation of traffic established by the IMO in the *Strait of Georgia* consists of a series of traffic separation schemes, two-way traffic lanes, and precautionary areas designed as follows (COLREG.2/Circ.51 and Circ.55):

(a) A precautionary area "GS" is established bounded by a line connecting the following geographical points:

(45) 48°52′.30 N, 123°07′.44 W
(46) 48°54′.81 N, 123°03′.66 W
(47) 48°49′.49 N, 122°54′.24 W
(48) 48°47′.93 N, 122°57′.12 W
(40) 48°47′.78 N, 122°59′.12 W
(39) 48°48′.19 N, 123°00′.84 W
thence to the point of origin (45).

(b) A separation zone in established bounded by a line connecting the following geographical positions:

(49) 48°53′.89 N, 123°05′.04 W
(50) 48°56′.82 N, 123°10′.08 W
(51) 48°56′.30 N, 123°10′.80 W
(52) 48°53′.39 N, 123°05′.70 W

(c) A traffic lane for north-westbound traffic is established between the separation zone and a line connecting the following geographical positions:

(46) 48°54′.81 N, 123°03′.66 W
(54) 48°57′.68 N, 123°08′.76 W

(d) A traffic lane for south-eastbound traffic is established between the separation zone and a line connecting the following geographical positions:

(53) 48°55′.34 N, 123°12′.30 W
(45) 48°52′.30 N, 123°07′.44 W

(e) A precautionary area "PR" is established bounded by a line connecting the following geographical points:

(53) 48°55′.34 N, 123°12′.30 W
(54) 48°57′.68 N, 123°08′.76 W
(55) 49°02′.20 N, 123°16′.28 W
(56) 49°00′.00 N, 123°19′.69 W
thence to the point of origin (53).

(f) A separation zone in established bounded by a line connecting the following geographical positions:

(57) 49°01′.39 N, 123°17′.53 W
(58) 49°03′.84 N, 123°21′.30 W
(59) 49°03′.24 N, 123°22′.41 W
(60) 49°00′.75 N, 123°18′.52 W
thence to the point of origin (57).

(g) A traffic lane for north-westbound traffic is established between the separation zone and a line connecting the following geographical positions:

(55) 49°02′.20 N, 123°16′.28 W
(62) 49°04′.52 N, 123°20′.04 W

(h) A traffic lane for south-eastbound traffic is established between the separation zone and a line connecting the following geographical positions:

(61) 49°02′.51 N, 123°23′.76 W
(56) 49°00′.00 N, 123°19′.69 W

5.4.7 Strait of Juan de Fuca

The IMO established two systems of separation of traffic and a precautionary area in the approaches to the Strait of *Juan de Fuca*, while it has four devices for the separation of traffic and another precautionary zone in the strait itself (COLREG.2/Circ.57).[379]

[379]In the *Puget Sound* – continuation of the *Strait of Juan de Fuca*, but in internal waters of the USA – the IMO adopted three devices for the separation of traffic in the *Rosario Strait*, in the access to the *Puget Sound* and in the strait itself (COLREG.2/Circ.51 y Circ.55).

Part I: Approaches to the Strait of Juan de Fuca

Western approach

(a) A separation zone is bounded by a line connecting the following geographical positions:

(1) 48°30′.10 N, 125°09′.00 W
(2) 48°30′.10 N, 125°04′.67 W
(3) 48°29′.11 N, 125°04′.67 W
(4) 48°29′.11 N, 125°09′.00 W

(b) A traffic lane for westbound traffic is established between the separation zone and a line connecting the following geographical positions:

(5) 48°32′.09 N, 125°04′.67 W
(6) 48°32′.09 N, 125°08′.98 W

(c) A traffic lane for eastbound traffic is established between the separation zone and a line connecting the following geographical positions:

(7) 48°27′.31 N, 125°09′.00 W
(8) 48°28′.13 N, 125°04′.67 W

South-western approach

(a) A separation zone is bounded by a line connecting the following geographical positions:

(10) 48°23′.99 N, 125°06′.54 W
(11) 48°27′.63 N, 125°03′.38 W
(12) 48°27′.14 N, 125°02′.08 W
(13) 48°23′.50 N, 125°05′.26 W

(b) A traffic lane for north-eastbound traffic is established between the separation zone and a line connecting the following geographical positions:

(14) 48°22′.55 N, 125°02′.80 W
(15) 48°26′.64 N, 125°00′.81 W

(c) A traffic lane for south-westbound traffic is established between the separation zone and a line connecting the following geographical positions:

(8) 48°28′.13 N, 125°04′.67 W
(9) 48°24′.94 N, 125°09′.00 W

Precautionary area "JF":

(5) 48°32′.09 N, 125°04′.67 W
(2) 48°30′.10 N, 125°04′.67 W
(3) 48°29′.11 N, 125°04′.67 W
(8) 48°28′.13 N, 125°04′.67 W
(11) 48°27′.63 N, 125°03′.38 W
(12) 48°27′.14 N, 125°02′.08 W
(15) 48°26′.64 N, 125°00′.81 W
(16) 48°28′.13 N, 124°57′.90 W

(18) 48°29′.11 N, 125°00′.00 W
(25) 48°30′.10 N, 125°00′.00 W
(17) 48°32′.09 N, 125°00′.00 W

thence back to the point of origin at (5).

Part II: Strait of Juan de Fuca

Western lanes

(a) A separation zone is bounded by a line connecting the following geographical positions:

(18) 48°29′.11 N, 125°00′.00 W
(19) 48°29′.11 N, 124°43′.78 W
(20) 48°13′.89 N, 123°54′.84 W
(21) 48°13′.89 N, 123°31′.98 W
(22) 48°14′.49 N, 123°31′.98 W
(23) 48°17′.02 N, 123°56′.46 W
(24) 48°30′.10 N, 124°43′.50 W
(25) 48°30′.10 N, 125°00′.00 W

(b) A traffic lane for north-westbound traffic is established between the separation zone and a line connecting the following geographical positions:

(26) 48°16′.45 N, 123°30′.42 W
(27) 48°15′.97 N, 123°33′.54 W
(28) 48°18′.00 N, 123°56′.07 W
(29) 48°32′.00 N, 124°46′.57 W
(30) 48°32′.09 N, 124°49′.90 W
(17) 48°32′.09 N, 125°00′.00 W

Traffic may exit the lane between points (29) and (30) or may remain in the lane between points (30) and (17) en route to the precautionary area.

(c) A traffic lane for south-eastbound traffic is established between the separation zone and a line connecting the following geographical positions:

(16) 48°28′.13 N, 124°57′.90 W
(31) 48°28′.13 N, 124°44′.07 W
(32) 48°12′.90 N, 123°55′.24 W
(33) 48°12′.94 N, 123°32′.89 W

Southern lanes

(a) A separation zone is bounded by a line connecting the following geographical positions:

(34) 48°10′.82 N, 123°25′.44 W
(35) 48°12′.38 N, 123°28′.68 W
(36) 48°12′.90 N, 123°28′.68 W
(37) 48°12′.84 N, 123°27′.46 W
(38) 48°10′.99 N, 123°24′.84 W

(b) A traffic lane for northbound traffic is established between the separation zone and a line connecting the following geographical positions:

(39) 48°11′.24 N, 123°23′.82 W
(40) 48°12′.72 N, 123°25′.34 W

(c) A traffic lane for southbound traffic is established between the separation zone and a line connecting the following geographical positions:

(33) 48°12′.94 N, 123°32′.89 W
(41) 48°09′.42 N, 123°24′.24 W

Northern lanes

(a) A separation zone is bounded by a line connecting the following geographical positions:

(42) 48°21′.15 N, 123°24′.83 W
(43) 48°16′.16 N, 123°28′.50 W
(44) 48°15′.77 N, 123°27′.18 W
(45) 48°20′.93 N, 123°24′.26 W

(b) A traffic lane for southbound traffic is established between the separation zone and a line connecting the following geographical positions:

(46) 48°21′.83 N, 123°25′.56 W
(26) 48°16′.45 N, 123°30′.42 W

(c) A traffic lane for northbound traffic is established between the separation zone and a line connecting the following geographical positions:

(47) 48°20′.93 N, 123°23′.22 W
(48) 48°15′.13 N, 123°25′.62 W

Eastern Lanes

(a) A separation zone is bounded by a line connecting the following geographical positions:

(49) 48°13′.22 N, 123°15′.91 W
(50) 48°14′.03 N, 123°25′.98 W
(51) 48°13′.54 N, 123°25′.86 W
(52) 48°12′.89 N, 123°16′.69 W

(b) A traffic lane for westbound traffic is established between the separation zone and a line connecting the following geographical positions:

(54) 48°14′.27 N, 123°13′.41 W
(55) 48°14′.05 N, 123°16′.08 W
(48) 48°15′.13 N, 123°25′.62 W

(c) A traffic lane for eastbound traffic is established between the separation zone and a line connecting the following geographical positions:

(40) 48°12′.72 N, 123°25′.34 W
(53) 48°12′.34 N, 123°18′.01 W

Precautionary area "PA", bounded by a line connecting the following geographical positions:

(33) 48°12′.94 N, 123°32′.89 W
(21) 48°13′.89 N, 123°31′.98 W
(22) 48°14′.49 N, 123°31′.98 W
(26) 48°16′.45 N, 123°30′.42 W
(43) 48°16′.16 N, 123°28′.50 W
(44) 48°15′.77 N, 123°27′.18 W
(48) 48°15′.13 N, 123°25′.62 W
(50) 48°14′.03 N, 123°25′.98 W
(51) 48°13′.54 N, 123°25′.86 W
(40) 48°12′.72 N, 123°25′.34 W
(37) 48°12′.84 N, 123°27′.46 W
(36) 48°12′.90 N, 123°28′.68 W

thence back to point of origin at (33).

5.4.8 Santa Barbara Channel

The IMO established a device for the separation of traffic which is divided into two parts (COLREG.2/Circ.24/Add1, March 10, 2000).

Part I: Between Point Vicente and Point Conception

(a) A separation zone bounded by a line connecting the following geographical positions:

(1) 34°20′.90 N, 120°30′.16 W
(2) 34°04′.00 N, 119°15′.96 W
(3) 33°44′.90 N, 118°35′.75 W
(4) 33°43′.20 N, 118°36′.95 W
(5) 34°02′.20 N, 119°17′.46 W
(6) 34°18′.90 N, 120°30′.96 W

(b) A traffic lane for north-westbound traffic is established between the separation zone and a line connecting the following geographical positions:

(7) 34°21′.80 N, 120°29′.96 W
(8) 34°04′.80 N, 119°15′.16 W
(9) 33°45′.80 N, 118°35′.15 W

(c) A traffic lane for south-eastbound traffic is established between the separation zone and a line connecting the following geographical positions:

(10) 33°42′.30 N, 118°37′.55 W
(11) 34°01′.40 N, 119°18′.26 W
(12) 34°18′.00 N, 120°31′.16 W

A safety fairway is established in the approach to Port Hueneme.

Part II: Between Point Conception and Point Arguello

(a) A separation zone bounded by a line connecting the following geographical positions:

(1) 34°20′.90 N, 120°30′.16 W
(6) 34°18′.90 N, 120°30′.96 W
(13) 34°23′.75 N, 120°51′.81 W
(14) 34°25′.70 N, 120°52′.51 W

(b) A traffic lane for westbound traffic is established between the separation zone and a line connecting the following geographical positions:

(7) 34°21′.80 N, 120°29′.96 W
(15) 34°26′.60 N, 120°51′.51 W

(c) A traffic lane for eastbound traffic is established between the separation zone and a line connecting the following geographical positions:

(12) 34°18′.00 N, 120°31′.16 W
(16) 34°22′.80 N, 120°52′.76 W

Chapter 6
The New Regime of Transit Passage Through International Straits

Transit passage through international straits is, undoubtedly, a neologism, an example of the new terminology introduced into International Law by the *United Nations Convention on the Law of the Sea* of 1982, and one of the most important achievements of the Grand Sea Powers in the new conventional regime.

Moreover, it is also new as a concept, which is evident if the rules of passage in transit of the UNCLOS – contained in the Second Section of Part III (articles 37–44) – are compared with the rules set out for straits in the 1958 Geneva Convention. The result of this comparison is that this regime is much more liberal than that of innocent passage and has strong connotations of the freedom in force on the high seas.

In synthesis the main differences between both systems of passage can be reduced to three. The first is that passage in transit includes the overflight of aircraft, which is probably the most problematic question, and innocent passage does not. The second is that passage in transit does not include the obligation which innocent passage has as concerns the fact that submarines must pass on the surface and flying their flag. Finally, although the power of the coastal State to adopt laws and regulations regarding passage in transit are specified in detail, it is easier to achieve this power than the similar power which the coastal State has in order to promulgate laws and regulations on innocent passage.

T. Scovazzi did not point out in vain that the new regime of passage in transit is so peculiar that the waters which are affected by this constitute "a *tertium genus*, distinct from both the internal waters and the territorial sea".[380]

According to article 37, this novel regime is applied to straits which are used for international navigation between one part of the high seas or an exclusive economic zone and another part of the high seas or an exclusive economic zone. As we have seen in Chap. 4, this takes in most of the international straits, and, undoubtedly, the straits which can be considered to be the most important in the world, from an economic-commercial point of view and from a strategic-military point of view,

[380] See Scovazzi (1995, p. 138).

A.G. López Martín, *International Straits*,
DOI 10.1007/978-3-642-12906-3_6,

that is to say, the main international navigation routes such as Gibraltar, Dover, Bab-el Mandeb, Hormuz, Malacca and Singapore.

We do not risk exaggerating when we state that this new regime has become the 'heart' of the regulation of the passage through the straits used for international navigation.

6.1 Conceptual Delimitation

Article 38 provides the meaning of the right of passage in transit and does so in the following terms:

> 2. Transit passage means the exercise in accordance with this Part of the freedom of navigation and overflight solely for the purpose of continuous and expeditious transit of the strait between one part of the high seas or an exclusive economic zone and another part of the high seas or an exclusive economic zone. However, the requirement of continuous and expeditious transit does not preclude passage through the strait for the purpose of entering, leaving or returning from a State bordering the strait, subject to the conditions of entry to that State.

This concept was proposed by the United Kingdom at the 1974 Caracas meeting, as a compromise formula between 'freedom of navigation' and 'innocent passage'. The essence of transit passage as distinct from innocent passage is precisely shown in the phrase "freedom of navigation and overflight solely for the purpose of continuous and expeditious transit".

A reading of this precept, classified by S.N. Nandan and D.H. Anderson as a "key provision in the Convention",[381] gives rise to several questions related to the exact determination of transit passage which we will attempt to analyse.

6.1.1 Beneficiaries of the Right to Transit Passage

The first clear conclusion which can be deduced from article 38 is that this is a right which is extensive to *ships* and *aircraft*. This conclusion is reinforced with the provisions in the first paragraph of article 38, which begins by stating that "in the straits referred to in the article 37, *all ships and aircraft enjoy the right of transit passage*, which shall not be impeded". A clear, unequivocal statement, with no exceptions, which confirms that transit passage does not distinguish nationalities or ownership, nor the type of ship or aircraft as it extends to *all*.

In fact, there is doctrinal unanimity as regards understanding that *all vessels* – including warships, nuclear driven vessels and those which transport harmful or hazardous substances – have the right to transit passage. On this point, the doubts

[381] See Nandan and Anderson (1989, p. 179).

we have observed in State practice regarding these types of vessels exercising innocent passage do not appear in the case of transit passage.

This is also a right which *all aircraft* have, an all embracing expression sufficient to include not only the civil aircraft used in air navigation, but also the other civil aircraft, military aircraft and other State aircraft. This last aspect is, undoubtedly, the most novel and certainly the most polemic of the new legal regime, as was shown in the course of the III Conference and as we established in Chap. 2 of this work. However, this was not due to civil air navigation, which had already been regularly overflying straits, but to military flights which had always required previous authorisation. Without doubt, this is the most outstanding difference in relation to the right of innocent passage.

Furthermore, as K.L. Koh points out, the right of transit passage "is not subject to the requirement of authorization or notification as a condition of passage".[382] A requirement which some States have imposed so that warships and, even, nuclear driven vessels or those which transport hazardous substances can exercise the right to innocent passage in their territorial waters, including straits as we have seen in the previous chapter.

However, there is one type of vessel which has given rise to doubts concerning how it must exercise transit passage; this is the *submarine* and other submergible vehicles. Concerning these, the interesting question arises regarding whether they can pass submerged or should pass on the surface.[383] This problem does not arise in innocent passage as the Convention is very clear in this respect; article 20 prohibits submerged navigation, and obliges submarines and any other submergible vehicles to navigate on the surface and fly their flags. However, there is no mention in any of the articles of Part III concerning submarines, thus it can be said that the Convention does not clearly and unequivocally guarantee the right of submerged transit.[384] However, this statement can be qualified, by endeavouring to interpret the whole of Part III which leads us to the conclusion that this right exists.

On the one hand, there is no similar provision to article 20, which expressly excludes submerged transit, and can be interpreted *sensu contrario* as permission since, if the drafters had wanted to make an exception of this, they would have incorporated a norm similar to innocent passage. On the other hand, article 38.2 uses the expression "freedom of navigation and overflight", which is situated within the high seas regime, therefore, it must be classified in association with what happens on the high seas; no provision expressly refers to submarines having this freedom to navigate submerged, however, there is no doubt that submarines have this freedom. Therefore, through similarity of reasoning, it seems to be pertinent that this right also corresponds to them under the transit passage regime as there is "freedom of navigation" although this is for the purposes of fast uninterrupted

[382]See Koh (1982, p. 153).

[383]For a detailed analysis of this problem, see the work of Burke (1976–1977, pp. 193–225).

[384]The contrary position to the recognition of a right of submerged transit passage is held fundamentally by H.G. Knight and W.M. Reisman (cited by Burke 1976–1977, p. 200).

transit. In addition, subsection (c) of article 39.1 refers to the "normal modes of continuous and expeditious transit". What constitutes a normal mode of navigation will depend on the type of vessel in question and, evidently, the normal mode of transit of a submarine is to navigate submerged. We must add to these arguments that there is "no interpretation to the contrary that has been advanced by any nation participating at UNCLOS".[385]

Furthermore, we must point out regarding this point that there is a difference in the drafting of article 38 and the corresponding norm of innocent passage as article 17 refers to "ships of all States", while article 38 omits the reference "of all States". A literal interpretation of this final provision implies that the holder of this right is the vessel and the aircraft, and not the flag or Registration State. Undoubtedly, the formulation is inadequate from the legal technical point of view as the vessel or aircraft cannot assume the obligations and responsibilities which are derived from the right granted.

6.1.2 Objective of This Passage Modality

Transit passage is understood solely as the exercise of the freedom of navigation and overflight for the purposes of expeditious and continuous passage as stated in article 38.2. Two conclusions can be deduced from this expression. The first is that this precision is intended to show that the vessels and aircraft only have the rights inherent to the mere continuous and expeditious passage through the strait, but not to the other rights deriving from the freedom of navigation and overflight, such as the exercise of military activity.[386]

The second involves the intention of the navigator to cross the strait continuously and expeditiously. This means that lingering, stopping or carrying out manoeuvres, which can be carried out on the high seas, are actions which are forbidden when the right of passage in transit is exercised.[387] In this regard, we can state that there is a certain parallelism between this provision and article 18.2, as article 18.2 classifies (innocent) passage as "continuous and expeditious". This language is repeated in article 53, as regards the right of passage through archipelagic sea lanes.[388] Article 18.2 permits stopping and anchoring as normal incidents of navigation, or in cases of force majeure or extreme danger. These last circumstances are

[385] Cf. Moore (1980, p. 121).

[386] See Reisman (1980, p. 72).

[387] Cf. Nandan and Anderson (1989, p. 181).

[388] Article 53 lays down the following:

> 1. An archipelagic State may designate sea lanes and air routes thereabove, suitable for the **continuous** and **expeditious** passage of foreign ships and aircraft through or over its archipelagic waters and the adjacent territorial sea.
> 2. All ships and aircraft enjoy the right of archipelagic sea lanes passage in such sea lanes and air routes.

also stipulated in transit passage in article 39.1(c), which provides that the vessels and aircraft "will abstain from all activity which is not related to its normal modalities of continuous and expeditious transit unless this is necessary due to force majeure or serious difficulties".

There is also a substantial difference between these two types of passage. Innocent passage is typified in article 19, based on 12 activities which are not innocent, which also permits the coastal State to perfectly identify any passage which is not innocent and impede this. No article defines nor typifies the meaning of "in transit", there is no criteria similar to 'innocence', which permits the coastal State to prevent passage not adapted to this as article 38 is silent on activities which are not adapted to transit.

A final precision must be made as regards the second paragraph of article 38, in the sense that the requirement for continual and expeditious transit will not impede passage through the strait in order to enter a coastal State of the strait, to exit this State or to return to it, subject to the conditions which regulate the entry to this State.

The United Kingdom introduced this point, which was defended before the Second Committee by its representative in the following terms:

> This delegation also had in mind the situation of the long strait which had more than one country bordering one side of the strait. Assuming a strait which had two countries on the western side, States A and B, and one country on the eastern side, State C, the United Kingdom draft proposed first, a right of transit should the ship or aircraft be going all the way northwards or southwards through the strait; secondly, a right of transit if the ship or aircraft was proceeding down the first part of the strait between States A and C with a view to calling at a port or airport of State B.[389]

Undoubtedly, an example of a strait which complies with the circumstances stated by the British delegate is the *Malacca Strait*, with Singapore as the State B in this equation, which has the strait itself as the sole access to and from its ports. Thus, the provision receives the name, "Singapore Clause".

Evidently, this entails an extension of the right of transit passage, as this can be exercised to enter a coastal State or to leave it. Therefore, from the meaning of article 38, it is clear that transit passage involves both *lateral and transversal passage*, that is to say, crossing the strait proceeding from the high seas or an exclusive economic zone, or *vertical passage*, that is to say, passage involving entering, leaving or returning to the interior waters of a coastal State of a strait. In this second case, as pointed out by D. Pharand,[390] the corresponding transit passage and not innocent passage will be applied, although passage will be subordinated to the conditions of entry or exit established by the coastal State. We disagree on this

---

3. Archipelagic sea lanes passage means the exercise in accordance with this Convention of the rights of navigation and overflight in the normal mode solely for the purpose of **continuous**, **expeditious** and unobstructed transit between one part of the high seas or an exclusive economic zone and another part of the high seas or an exclusive economic zone (the letters ain bold are ours).

[389] See Second Committee, 3rd meeting (1974), p. 32, *II Off. Rec.*, 125.

[390] See Pharand (1977, p. 82).

point as regards the opinions of such as V. Bou Franch,[391] who believes that transit passage is applied to transversal navigation, and is excluded in the case of perpendicular or vertical navigation.

6.2 Reciprocal Rights and Obligations of the Coastal States and the Users of the Straits Regulated by Transit Passage

The rights and obligations which fall on the coastal States, as well as on the vessels and aircraft which use these international straits regulated by transit passage, are specifically regulated in articles 39–44 of the UNCLOS. Following the structure marked out by the Convention, we will distinguish the obligations of the vessels and aircraft during their transit passage and the rights of the coastal States, and finally, the common obligations of both.

6.2.1 Obligations of Vessels and Aircraft During Transit Passage

This question is regulated in article 39, which establishes obligations which are generic for vessels and aircraft, consequently, these apply to the respective flag or registration states of the aircraft, and others are specific for the transit passage of each of these.

We should point out that the obligations listed in this provision are obligations which arise from the 1982 Convention for the vessels and aircraft which exercise the right of transit passage. These are, as stated by H. Caminos,[392] obligations which arise from the 1982 Convention for the vessels and aircraft which exercise the right of transit passage, and are different from the additional obligations arising as a consequence of the implementation of sea lanes and schemes for the separation of traffic by the coastal State and the regulatory powers attributed to this as regards transit passage. As we have stated above, the correlative of a right is an obligation.

6.2.1.1 Common Obligations for Ships and Aircraft

The obligations which are common to ships and aircraft are determined in the first paragraph of article 39, a provision which is, undoubtedly, clearly associated to the concept of transit passage. These are the following:

> Ships and aircraft, while exercising the right of transit passage, shall:
> (a) proceed without delay through or over the strait;

[391] Cf. Bou Franch (1994, p. 84).

[392] See Caminos (1987, p. 148).

(b) refrain from any threat or use of force against the sovereignty, territorial integrity or political independence of States bordering the strait, or in any other manner in violation of the principles of international law embodied in the Charter of the United Nations;
(c) refrain from any activities other than those incident to their normal modes of continuous and expeditious transit unless rendered necessary by *force majeure* or by distress;
(d) comply with other relevant provisions of this Part.

As stated in the introductory phrase to this first paragraph, the obligations set out are applied when the ships or aircraft are exercising the right of transit in passage. In this regard, we must take into consideration that these ships or aircraft exercise this right when they enter a strait with the intention to pass through it, or to cross it in order to exit or head for a port or airport of a coastal State of the strait.

The first obligation which falls on a ship or aircraft which exercises the right of transit passage through a strait is that of "*proceed without delay*". This expression does not have an identical meaning but is in consonance with the requirement involving continuous and expeditious transit imposed by articles 38.2 and 39.1(c). In this same regard, it is understood that ships must move forward at their normal speed, taking into account all the relevant factors, including the safety requirements, the meteorological conditions or the presence of other vessels or aircraft in the straits.[393]

The second obligation refers to the abstention from *any threat or use of force*, using an identical formulation to that used by article 19.2 concerning innocent passage, although, in this last case, this is completed with a number of activities considered to be harmful to peace, good order and security in the coastal State, which does not take place in transit passage; such as the use of weapons, the obtaining of information propaganda, the take off and landing of aircraft or military schemes. Faced with this divergence, the doubt arises whether these activities are allowed or not. According to D. Pharand,[394] as they are not expressly prohibited, as in innocent passage, they could be considered to be permitted in transit passage. We do not share this opinion. If we take into account that an activity included in article 19.2 which is not mentioned in Part Parte III is that concerning the disturbance of the communication systems or any other coastal installation or service, and these activities infringe the property rights and the exclusive jurisdiction of the coastal State, does this mean that these are permitted? Evidently not, this would entail a clear infringement of the sovereignty of the coastal State. As regards this point, it should be borne in mind that Part III does not limit the authority of the coastal State of the strait as regards the activities related to transit passage. Article 34 safeguards the exercise of sovereignty and jurisdiction of the coastal State over the waters of the strait, the airspace and soil and subsoil, and, undoubtedly, the competence to carry out these activities is exclusive to the coastal State. This conclusion is also strengthened by what is stipulated in article 38.3, in the sense that any activity

[393]Cf. Nandan and Anderson (1989, p. 183).

[394]See Pharand (1977, p. 83).

which is not an exercise of the right of transit passage through a strait remains subject to the other applicable provisions of this Convention.

Furthermore, none of the activities mentioned in article 19.2 can be considered to be an activity related to the normal modalities of fast, expeditious transit, as required b subsection (c).

The third obligation involves refraining *from any activities other than those incidents to their normal modes of continuous and expeditious transit*. The text does not define the normal modalities of transit, but it can be deduced that vessels can carry out the common and usual activities involved in navigation, depending on their type and the hydrographical and navigational characteristics of the strait through which they are passing, which will include the submerged passage of submarines, as we pointed out above. In the case of aircraft, we can also consider that this refers to the usual modes of navigation for each type of aircraft, while in the case of hydrofoils and hovercraft, this involves flying or sliding over an air cushion.

However, this obligation is excluded in the case of force majeure, for example, a collision or hurricane or serious difficulty. The meaning of this subsection (c) is similar to the meaning used in article 18.2, but not identical as there is quite a difference. Article 18.2 also recognises the exception involved in providing assistance to persons, vessels or aircraft in danger or in serious difficulties. In its amendment to article 39, Spain[395] proposed that this exception included in innocent passage be incorporated, but it was rejected and is not included as such. One interpretation *sensu contrario* of article 39.1(c), would mean that, if a vessel or aircraft in transit through a strait carries out an action involving humanitarian aid, assistance to persons, vessels or aircraft, it would lose the right of transit passage as it would not be complying with this. This is absurd if we take into account that the tradition to provide assistance at sea is as ancient as navigation itself.

Finally, article 39.1 refers to the *duty to comply with other relevant provisions of this Part*.

Specifically, in the first place, this involves the prohibition involving foreign vessels, included in article 40 in the following terms: "During transit passage, foreign ships, including marine scientific research and hydrographical survey ships, may not carry out any research or survey activities without the prior authorization of the States bordering straits". It is important to remember that, according to article 40, the prohibition is not restricted to 'marine' research, nor to 'hydrographical' surveys, but this involves a general prohibition concerning any kind of research or survey carried out by vessels while they are in transit passage. Thus, the exclusive competence of the coastal State in this matter is recognised, as occurs with innocent passage; and we must also place this regulation in relation to article 245, which grants the coastal State to regulate, authorise and carry out scientific research in its territorial sea.

Furthermore, the express reference to vessels engaged in scientific marine research and hydrographical surveys must not be seen as a redundancy, but as a

[395]Cf. Doc. A/CONF.62/L.109 (1982).

clarification intended to prevent confusion or erroneous interpretations. If article 39.1(c) regarding the normal modalities of transit is taken into account, there is no doubt that the normal modality of these vessels is research and surveys, thus, this clarification which excludes these vessels from this activity without authorisation is necessary. We must also draw attention to the use of the plural "States", which means that the consent of all the coastal States of the strait is required, obviously when there is more than one coastal State, in order to carry out activities involving research or surveys. If only the consent of one of these coastal States is obtained, these activities can only be carried out in the part of the territorial sea which corresponds to this State.

During transit passage, the vessels must respect the sea lanes and the traffic separation schemes designated by the coastal State of the strait and adopted by the IMO, in accordance with the stipulations in article 41.7. These vessels must comply with the laws and regulations dictated by the coastal State as regards transit passage, as stipulated in article 42.4.

However, none of these three provisions affects aircraft as they only refer to ships, which demonstrates a lack of judicial coherence, and shows how those drafting the Convection extended the obligations for ships "ad sideram" simply by inserting the words 'aircraft' and 'overflight' in the articles corresponding to sea navigation, and does not take the trouble to examine the consequences nor carry out the adaptations required.

6.2.1.2 Exclusive Obligations for Vessels

Ships in transit passage shall comply with generally accepted international regulations, procedures and practices for safety at sea, including the *International Regulations for Preventing Collisions at Sea*, known as COLREG. This also involved the generally accepted international regulations, procedures and practices for the prevention, reduction and control of pollution from ships (article 39.2).

The expression "generally accepted" used in this article was drafted in intentionally broad terms in order to include the international agreements on this matter and which the States are parties to, such as the customary regulations generally accepted in these matters and the agreements which include these. The 'generally accepted' condition was suggested by the United States in the III Conference in order to prevent the imposition of unilateral criteria by the coastal State.[396]

Undoubtedly, the main international instruments as regards the safety of navigation, as well as the prevention and control of pollution, have been drafted within the framework of the IMO, which is the international organisation competent in this area (with 169 Contracting States). In addition to the aforementioned COLREG of 1972 (which has 153 Party States), mention should be made of the *International Convention for the Prevention of Pollution from Ships*, MARPOL 73/78 (with

[396]Cf. Doc. A/AC.138/SC.II/SR.33-47.

150 party States to Annex I/II, 133 to Annex III, 124 to Annex IV, and 139 to Annex V) and its 1997 Protocol and Annex VI (with 58), and the 1974 *International Convention for the Safety of Life at Sea* (SOLAS) (with 159 Party States) and its two Protocols of 1978 (with 114 party States) and 1988 (with 94), as the main referents.[397]

6.2.1.3 Obligations Applicable Only to Aircraft

Similar to the above, the third paragraph specifies a number of duties which exclusively affect the aircraft which fly over the straits used for international navigation, exercising their right to transit passage; obligations which are intended to ensure the safety of air navigation:

> Aircraft in transit passage shall:
> (a) observe the Rules of the Air established by the International Civil Aviation Organization as they apply to civil aircraft; state aircraft will normally comply with such safety measures and will at all times operate with due regard for the safety of navigation;
> (b) at all times monitor the radio frequency assigned by the competent internationally designated air traffic control authority or the appropriate international distress radio frequency.

It is evident that the primordial obligation of civil aircraft is to respect the *Rules of the Air* of the International Civil Aviation Organization (ICAO). We should mention that these Rules are contained in an Annex to the Chicago Convention of 1944 and contain international standards and practice recommended as regards safety and air traffic obligatory in the air apace above and adjacent to the high seas. Thus, what article 39.3 does is convert the rules of the high seas, into obligatory for the overflight of straits regulated by transit passage, which we must not forget continue to be territorial sea, and this circumstance again corroborates the idea which we have so often repeated that the new regime of transit passage equates the waters of international straits with the high seas as regards the navigation regime.

[397]Status at March 1, 2010. The status of the Conventions held within the framework of the IMO can be seen at http://www.imo.org/Conventions/mainframe.asp?topic_id=247. Moreover, we must point out that there are more than 40 agreements which attempt to regulate the safety of navigation and the prevention of pollution; as examples, we can cite the following: the *Convention on the Prevention of Marine Pollution by Dumping of Wastes and Other Matters* (LCD), 1972; *International Convention on the Control of Harmful Anti-fouling Systems on Ships* (AFS), 2001; *International Convention for the Control and Management of Ships' Ballast Water and Sediments*, 2004; *International Convention on Civil Liability for Oil Pollution Damage* (CLC), 1969; *International Convention on the Establishment of an International Fund for Compensation for Oil Pollution Damage* (FUND), 1971, *Convention relating to Civil Liability in the Field of Maritime Carriage of Nuclear Material* (NUCLEAR), 1971; *International Convention on Civil Liability for Bunker Oil Pollution Damage*, 2001. These texts can be seen at http://www.imo.org/Conventions/mainfarme.asp?topic_id=260.

Nevertheless, it is not clear to what extent and under what conditions this same obligation binds State aircraft due to the introduction of the adverb "normally" in order to specify this obligation. In fact, the Convention does not clarify what must be understood as 'normally', nor who must determine when the situation is 'normal' or 'abnormal', so, it can be easily concluded that this would be the State where the aircraft is registered, endorsing the effective conduct of the pilot, who would finally be empowered to decide, at his own discretion, whether a determined situation comes under the category of 'abnormal' in order to be 'released' from compliance with the Rules of the Air, included in the requirements for identification.

Undoubtedly, this is a crucial question as the inevitable consequence of this specification, interpreted *sensu contrario*, is that, in abnormal circumstances, the aircraft State is exempt from the obligation to comply with these Rules of the Air. This issue led to the drafting of the absurd article 39.3 which is a contradiction, since, if the finality of this provision is to guarantee the safety of navigation, which is considered of supreme interest, why is this relegated and excluded by the mere classification of a situation as abnormal? The meaning of this norm leads us to doubt the authentic 'supreme' interest of transit passage of State aircraft through international straits as, evidently, this is not the safety of navigation but safeguarding the particular interests of the main sea powers, and we know who these are as we should remember that military aircraft are included in the State aircraft.

As the Spanish representative at the III Conference correctly pointed out, "it is evident that establishing an obligation which will only have to be complied with is the same as not establishing any obligation".[398] The mere classification of a situation as 'abnormal' sets these aircraft aside from all the regulations, which represents a threat to air navigation, to the safety of the coastal States and their populations. This particular situation, which is generated by the new regulations, meant that Spain submitted an amendment to article 39, in the sense that it suppressed the term "normally". However, the amendment was rejected by 55 votes against and 60 abstentions.[399]

Undoubtedly, the key question concerning this obligation is the 'classification' of a situation as normal. The interpretation declarations made by Spain, both on signing and ratifying the UNCLOS, are aimed at clarifying this controversial point, and understand that the word "normally" means "except in the case of force majeure or serious difficulty".[400] The concern behind the formulation of this declaration is none other than the attempt to attenuate the problems which presumably will derive from the wide discretional margin granted to the aircraft of a State when theses are in transit through the straits, when it is precisely the indiscriminate passage of these types of aircraft which might cause the greatest inconvenience to the sovereignty of the coastal States of the straits. Although we must not forget that

[398]See Intervention of the Ambassador J.M. Lacleta Muñoz, in Doc. A/CONF.62/SR.182, of April 30, 1982.

[399]Cf. Doc. A/CONF.62/SR.176.

[400]Cf. http://untreaty.un.org.

this is a mere interpretative declaration which has no judicial effects unless this is expressly accepted by the other States which are parties to the Convention, which is quite improbable.

The second obligation imposed by article 39.3(b) on aircraft refers to the *rules of communication*, specifically, to the duty to keep the radio frequency assigned by the competent internationally designated air traffic control authority, or the corresponding international assistance radio frequency tuned in at all times. We must draw attention to the fact that the use of the disjunctive *"or"*, which establishes an option right for the aircraft which may choose between one system of communication or another with no need to provide a justifying reason. The use of this disjunctive conjunction is a new concession to the aircraft which carry out transit passage, to the detriment of the State being flown over, and this also involves a problem concerning the application of successive treaties regarding this matter as Annex 10 of the Chicago Convention obliges aircraft to carry two frequency systems. In this regard, article 39 should have maintained the cumulative obligation to tune into the two radiofrequencies, in line with the Chicago Convention, and not the option, but this has not been so.

Furthermore, this provision provides an incomplete reproduction of the rule in Annex 2, according to which, an aircraft operated as a controlled flight must stay tuned into the proper radiofrequency of the competent traffic control unit at all times, and establish the necessary communication and dialogue with this unit (rule 3.6.5.1). This last detail is not incorporated into article 39.3(b). The obligation to simply stay tuned into the radiofrequency assigned by the control tower of the State being flown over is of little use if the pilot does not maintain a dialogue with the air traffic controllers and does not follow their instructions.

In any case, as pointed out by the Report drafted by the ICAO in 1984,[401] this is an error which has continued from the time it was initially drafted in 1977, and has not been corrected despite the notification of the ICAO made to the Draft Committee of the Secretariat of the United Nations. It adds that, in practice, this is a matter of minor importance and does not entail a conflict with the rules of the ICAO, as these constitute *lex specialis* which must be applied in the case of the Convention.

However, we disagree with this 'optimistic' view of the ICAO. As we stated above, we are again facing a problem concerning the application of successive treaties relating to the same subject matter, whose rules of application are established in article 30 of the 1969 *Vienna Convention on the Law of Treaties*. None of the criteria involved gives preference to the *lex specialis*, nor is the special norm very clear on this point as the UNCLOS could be considered to be a *lex* specialis as regards the regime of overflight of straits. According to article 30, it would be necessary to examine whether there is a relationship clause in either of the treaties. In this regard, both Conventions have a relationship clause, which are not clauses of subordination or compatibility, which solve this problem, but both are primacy clauses. In fact, both article 82 of the Chicago Convention, and article 311.2 of the

[401]Cf. Doc. ICAO/.C/WP/7777, January 20, 1984, pp. 9–12.

UNCLOS, only allow agreements which are compatible with these, thus, the problem is not resolved. Taking into account the fact that there is no identity of parties as regards both Conventions, the solution to this question is regulated in paragraph 4 of article 30 of the 1969 Vienna Convention, as follows: as concerns the relationships between States parties to both treaties, the latter treaty is applied; in the relationships between States which are parties to the two treaties, the treaty to which both States are parties will be applied, regardless of whether this is the anterior or the posterior treaty. Evidently, as regards the relationships between States which are parties to one treaty, this treaty will be applicable.

The Chicago Convention has 190 Contracting States and the UNCLOS has 160 Contracting States (both at March 1, 2010). There are 36 States which are parties to the Chicago Convention and not to the UNCLOS. Obviously, these States apply the provisions of the 1944 Convention, just as the States which are parties to the Chicago Convention and these 36 States must apply the provisions of the Chicago Convention. Attention must be drawn to four of these 36 States[402] as these are coastal States of straits governed by transit passage. Such is the case of Iran and the United Arab Emirates, coastal States of the *Strait of Hormuz*. The United States of America, coastal State of the Straits of *Bering, Agattu, Tanaga, Unimak, Samaiga, Seguan, Adak, Kaulakhi, Kaiwi, Pailolo, Alalakeiki, Kealaikahiki, Auan* and *Kalohi*. And Venezuela, the coastal State of the *Aruba-Paraguana Passage, Serpent's Mouth* and *Dragon's Mouth*. This means that, in the straits we have referred to the territorial seas belong to these States and overflying aircraft must follow the obligations established by the 1944 Chicago Convention.

There are five States which are parties to the UNCLOS but are not parties to the Chicago Convention. In this case, the solution is the same but inversely, that is to say, as regards the relationships between them and their relationships with States which are parties to both Conventions, the provisions we have seen in the UNCLOS will be applied. We must draw attention to one of these five States which is a coastal State of two article 37 international straits, this is Dominica, and the straits are the *Dominica Passage* and the *Martinique Passage*. Undoubtedly, the obligations imposed by the 1982 Convention govern the use of these straits.

The other 153 States which are parties to both Conventions must apply the provisions of the UNCLOS as regards their mutual relationships as this is the later treaty. This is apart from the questions of international liability which might arise from non-compliance with the obligations of any of the Conventions,[403] in the sense stated in article 30.5 of the 1969 *Convention on the Law of Treaties*.

---

[402]The other 32 States which are not parties to the UNCLOS are parties to the Chicago Convention are: Afghanistan, Andorra, Azerbaijan, Bhutan, Burundi, Cambodia, the Central African Republic, Colombia, Congo, the Democratic Republic of Korea, Ecuador, El Salvador, Eritrea, Ethiopia, Israel, Kazakhstan, Kyrgyzstan, Lesotho, Liberia, Libya, Malawi, Niger, Peru, Ruanda, San Marino, Swaziland, Syria, Tajikistan, Thailand, East Timor, Turkey and Turkmenistan.

[403]In fact, article 82 of the Chicago Convention obliges the contracting States not to contract obligations and mutual understandings which might be incompatible with the provisions of the Convention.

This is not a minor question, but it may entail serious consequences in the case of an aircraft which enters the airspace of a State since, as stated in the 1978 Spanish Memorandum, "if this aircraft fails to maintain radio contact with the air traffic control tower and fails to obey its instructions, it will be a real hazard for civil aviation; moreover, it would also be impossible to distinguish an aircraft which wishes to exercise the right of transit passage from a possible aggressor aircraft if there is no previous identification".[404]

Furthermore, given the fact that this obligation to maintain the radiofrequency tuned in, as stated in article 39.3(b) which repeats Annex 2 of the Chicago Convention, is intended for **controlled** flights, it must be interpreted in the sense that **uncontrolled** flights over straits are forbidden.

We must mention that the provisions in articles 39 and 40 related to the obligations of ships and aircraft which exercise the right of transit passage through article 37 straits, is applicable, "mutatis mutandi", for the ships and aircraft which navigate or fly over 'archipelagic straits' designated by the archipelagic State as archipelagic sea lanes in the sense of article 53 (article 54) UNCLOS.[405]

One final note concerning the regulation analysed is that we wish to stress the fact that in the second section and in the third section of article 39, the obligations of the ships and aircraft mentioned arise in reference or in relation to the coastal State or States of the strait, but with regard to an international organisation, such as the IMO and the ICAO. This situation proves to what extent the interests of States other than the coastal States have been taken into account have been taken into account in the new transit passage regime.

6.2.2 Rights of the Coastal State or States of an International Strait

As we have already analysed, unlike the innocent passage regime, the UNCLOS establishes a number of normative powers for the coastal State, such as the consequent jurisdictional and protection competences, as regards transit passage, the Convention seems to completely ignore competences. In fact, articles 42 and 41 of the UNCLOS establish the regulatory powers the coastal States have as concerns transit passage through the straits, but they omit any reference to the corresponding executive competences in this regard, which is another indication that the right of transit passage seems to be above the sovereignty of the coastal States for the 1982 Convention.

This is an evident omission which we must complete by carrying out the due interpretation of the provisions of the Convention.

---

[404]See the reference in Yturriaga Barberán (1991b, pp. 414–415).

[405]For a comparative analysis of the rights and obligations which affect the aircraft in transit over straits or archipelagic waters, see Vries Lentsch (1983, pp. 165–225).

6.2.2.1 The Regulatory Competence of the Coastal States Concerning Transit Passage Through Their Straits

The regulatory competences of the coastal States of international straits essentially consist of adopting laws and regulations related to transit passage through their straits (article 42), and designating sea lanes and prescribing schemes for the separation of traffic in these lanes (article 41). We must draw attention to the fact that these rights recognised for the coastal State are quite limited as it can only dictate laws and regulations in four specific material areas, and establishing the sea lanes and the schemes for the separation of the traffic requires the agreement of the IMO. In addition, all these refer to sea navigation and not to air navigation; even the establishment of lanes and schemes for the separation of traffic is limited only to sea navigation, and there are no regulations concerning similar measures related to air traffic.

In fact, article 42 establishes that the coastal States of straits can adopt laws and regulations related to transit passage especially or to some matters which are on a limited list, and which refer to maritime safety, pollution, fishing and customs, tax, health and immigration laws. A list which contains important omissions, such as liability for damages, and which contrasts with the regulation powers recognised in article 21.1 for the coastal States governed by innocent passage. This entails a substantial restriction to the powers granted to the coastal State as regards ordering passage.

A restriction which is intensified if we observe how the regulatory rights included in articles 22 and 23 concerning innocent passage by nuclear driven vessels or those which transport harmful or hazardous substances through the straits governed for the article 45, do not exist for the transit passage of these types of vessels with special characteristics through straits of the article 37. Not even the possibility to charge levies as remuneration for certain services provided for ships which cross the straits in transit, recognised in innocent passage (article 26), is a right granted to the coastal States in the case of transit passage through their straits.

The regulatory powers of the coastal States concerning transit passage are extended to the stipulations in article 42.1, each of whose four sections specifies the subject of the legislative competence of the coastal State and restricts its scope in the following terms:

Subject to the provisions of this section, States bordering straits may adopt laws and regulations relating to transit passage through straits, in respect of all or any of the following:

- *"(a) the safety of navigation and the regulation of maritime traffic, as provided in article 41"*

Unlike article 21 which we have analysed and which authorised the coastal State to dictate norms only concerning the safety of navigation and the regulation of sea traffic, article 42 qualifies this power and refers to article 41, that is to say, it limits this power to the mere possibility to designate sea lanes and schemes for traffic safety, an important question which will analysed in detail in the following section due to the importance of the matter.

• *"(b) the prevention, reduction and control of pollution, by giving effect to applicable international regulations regarding the discharge of oil, oily wastes and other noxious substances in the strait"*

The rigidity and restrictive nature of the terms of this provision contrast with the parallel provision of article 21.1 concerning innocent passage, where the limitations we observe here do not appear. In fact, the legislative competence of the coastal State regarding the prevention, reduction and control of pollution in the straits where transit passage is in force is not absolute, but has two important limitations. One is that its laws and regulations must have the finality that they provide effect to the international regulations applicable. Another is that it only restricts the unloading of hydrocarbons, waste oil and other harmful substances; an absurd limitation which excludes the regulations on pollution due to other reasons, such as spills or the pollution resulting from maritime accidents.

Spain proposed two amendments in order to overcome these limitations, one tended to replace the word "applicable" by "generally accepted", and the other to eliminate "oil".[406] The second was withdrawn, but the first was maintained and accepted by a margin of 60 votes in favour, 29 against and 51 abstentions. However, despite its acceptance, the rigid Conference Rules prevented its incorporation to the text of the UNCLOS. The difference in meaning is important because the expression "generally accepted" includes customary norms, incorporated or not into an international treaty, accepted by the majority of States; while "international regulations applicable" is limited to international treaties in force which the coastal State of the strait and the flag State of the vessel in transit. Among these treaties we must cite MARPOL 73/78.

As pointed out by T. Scovazzi[407] it is surprising that the coastal State is not also granted the right to know whether the vessels in transit through their straits carry contaminating substances.

Exercising this power and as a partial incorporation of Directive 2005/35/EC of the European parliament and of the Council, of September 7, 2005, Spain adopted Royal Decree 394/2007, of March 31, *on measures applicable to ships in transit which unload pollutants in Spanish marine waters*, which will be applied to the unloading of contaminating substances in Spanish straits used for international navigation subject to transit passage, as expressly set out in paragraph 2 of its article 2. Among other measures, in the event that a vessel which is navigating in the territorial sea has unloaded pollutant which supposes or might suppose considerable damage to the natural resources of these waters or to the coast or property linked to it, article 3 of this Royal Decree legitimises the Spanish Maritime Administration to adopt the necessary policing, including detaining the vessel, as well as the possible commencement of the sanctioning proceedings or the referral of

[406]Cf. Doc. A/CONF.62/L.123, of April 13, 1982.

[407]See Scovazzi (1995, p. 142).

the actions to the Prosecutor.[408] This legislation is in force in the *Strait of Gibraltar* and in the *Menorca Channel*, straits where Spain is a coastal State.

• *(c) with respect to fishing vessels, the prevention of fishing, including the stowage of fishing gear*

Again the regulatory competence of the coastal State regarding this matter is much more limited in the straits where transit passage is in force than in those with innocent passage. In the latter, the coastal State can dictate laws and regulations for the conservation of the living resources of the sea and to the prevention of infringement of the fisheries laws and regulations of the coastal State (article 21.1); as regards transit passage, it can only regulate the prohibition to fish. This is a restrictive and unjustified formulation, which contradicts the stipulations in article 34 which declares that the regime of passage through straits used for international navigation established in this Part shall not in other respects affect the legal status of the waters forming such straits or the exercise by the States bordering the straits of their sovereignty or jurisdiction over such waters and their air space, bed and subsoil

What happens when the natural resources are not fish? Can these be regulated in a strait, territorial sea, with transit passage in force?

The illogical nature of the terminology of this subsection goes to an extreme when it restricts the regulation to "fishing vessels". Can the coastal State regulate fishing if the vessels are not fishing vessels? Can it prohibit foreign vessels which fish in its territorial sea or must it restrict itself to fishing vessels?

• *"(d) the loading or unloading of any commodity, currency or person in contravention of the customs, fiscal, immigration or sanitary laws and regulations of States bordering straits"*

Of course, the regulations in this regard are more restricted than in innocent passage. Furthermore, they are more restricted than in the adjacent zone since the coastal State can regulate the infringement of its laws and regulations on taxation, Customs, health and immigration, as well as prevent violations of its regulations under the regime of innocent passage and in the adjacent zone. The coastal State of a strait does not have this power under the regime of transit passage.

There are 35 coastal States of straits where transit passage is in force[409] – status at March 1, 2010 –, none of these have dictated specific laws and regulations concerning this new legal regime of passage, except for the recent Spanish Royal Decree of March 31, 2007, on measures applicable to vessels in transit which unload pollutant in Spanish marine waters, which we have already referred to (A Royal Decree which, however, refers to all the Spanish marine waters and not

---

[408]The Spanish Ministry of Development must dictate the Implementation Rules for this Royal Decree. See *Boletín Oficial del Estado* No. 81, of April 4, 2007, pp. 14667–14669.

[409]These 35 States are: the United Kingdom, France, Spain, Morocco, Italy, Ukraine, Russia, Greece, Yemen, Djibouti, Somalia, Oman, the United Arab Emirates, Japan, China, Malaysia, Singapore, Indonesia, India, Sri Lanka, Australia, Papua New Guinea, Nueva Zealand, Samoa, the United States, Canada, Denmark, the Bahamas, Cuba, Dominica, St. Lucia, St Vincent, Trinidad and Tobago, Venezuela and Holland.

exclusively to straits). This means that the internal legislation in force continues to be the general legislation on territorial seas, which does not establish a differentiated regime for straits. This legislation was adequate until the corresponding entry into general and particular force of the UNCLOS, but now it is clearly insufficient.

The change became necessary due to the extent of the legal insecurity as regards the users of these international straits. The coastal States must make the necessary changes to their internal legislation in order to adapt it to the new transit passage regime which is in force in their straits.

In fact, only five States, Djibouti, Cuba, Spain, the United States and the United Kingdom have expressly recognised the new navigation regime in their domestic law. However, two of these have restricted this to a referral to the international rules in this matter, without specifying what these are. These are Djibouti, the coastal State of the *Strait of Bab-el-Mandeb*, and Cuba, which is the coastal State of the *Old Bahama Channel*. In fact, article 15 of Law No. 52/AN/78 of Djibouti, *on the territorial sea, adjacent zone, exclusive economic zone, maritime and fishing borders*, provides that "the provisions of this Law shall not modify the international rules of navigation in the Straits of Bab El Mandeb". In the same sense, article 5 of Decree-Law No. 1 of February 24, 1977 concerning the width of the territorial sea of the Republic of Cuba establishes that "The Republic of Cuba, without prejudice to the provisions of the preceding articles and in accordance with the principles of International Law, will respect the freedom of navigation and overflight as regards the straits or channels used for international navigation".[410] Only the United Kingdom, the United States of America, and recently Spain, have expressly recognised the application of transit passage for their straits. The United Kingdom in its *Territorial Sea Act* 1987[411] and in the *Air navigation Order* of November 29, 1989,[412] in relation to the *Straits of Dover* and *The Hole*. The United States declared in its Proclamation of December 27, 1988, that "within the territorial sea of the United States, the ships of all countries enjoy the right of innocent passage and the ships and aircraft of all countries enjoy the right of transit passage through international straits".[413] With regard to Spain, after its initial implacable opposition to transit passage, its position has become more flexible to the extent that article 2 of Royal Decree 394/2007, of March 31, refers expressly to "the straits used for international navigation subject to the *regime of transit passage* where Spain

[410]See http://www.un.org/Depts/los/LEGISLATIONANDTREATIES/regionalist.htm in order to see the texts of these two provisions of internal law.

[411]Cf. *Law of the Sea Bulletin*, No. 10, 1987, p. 11 et seq.

[412]This Order of 1989 recognises the right of transit for foreign aircraft which means, "overflight of the strait from an area of high seas at one end of the strait to an area of high seas at the other end, or flight to or from an area of high seas over some part of the strait for the purpose of entering, leaving or returning from a State bordering the strait and 'an area of high seas' means any area outside the territorial waters of any State" (see *BYBIL*, 1989, p. 668).

[413]See *ILM*, vol. 28, 1989, p. 284.

exercises jurisdiction";[414] this is the evident result of having backed down and ratified the 1997 UNCLOS.

The other 30 States have not mentioned the new regime of transit passage in their legislation at all. Moreover, there are States which have even expressly stated their intention to maintain the legislation in force previous to their joining the UNCLOS. This is the case of Morocco which, on ratifying the UNCLOS on May 31, 2007, stated that "the laws and regulations relating to maritime areas in force in Morocco shall remain applicable without prejudice to the provisions of the United Nations Convention on the Law of the Sea".[415] That is to say, despite ratifying the Convention, it maintains its internal legislation regarding marine areas in force which includes the Law of 1973, whose article 3 establishes the regime of innocent passage for the *Strait of Gibraltar*.

Moreover, we must bear in mind that, among these 30 States we have referred to, there are three States which are not parties to the UNCLOS and have also manifested their inconformity with this new regime: the United Arab Emirates, Iran (both coastal States of the *Strait of Hormuz*), and Venezuela, therefore, we cannot state that they are obliged by the new regime of passage.

In addition, we must point out that there are six coastal States of three straits of article 37, which have signed international treaties regarding these straits in which they have recognised the right of transit passage. This is the case of the Treaty made by Australia and Papua New Guinea on *sovereignty and marine borders* between the two countries, including the *Torres Strait*, of December 18, 1978, whose article 7 recognises the freedom of navigation and overflight in the *Torres Strait*. The *Treaty on the Border Delimitation* between Venezuela and Holland of March 31, 1978, whose article 4 recognises the existence of the right of transit passage between the Island of Aruba and the Venezuelan mainland, that is to say, in the *Aruba-Paraguana Passage*, although this applies only for their respective vessels and aircraft as article 4.3 excludes State aircraft and the warships of third party States. Finally, the Agreement of November 2, 1988 made between France and the United Kingdom *on the delimitation of their marine border*, which was accompanied by a Joint Declaration concerning the status of the *Dover Strait*, which stated that "the two Governments recognise rights of unimpeded passage for merchant vessels, state vessels and, in particular, warships following their normal mode of navigation, as well as a right of overflight for aircraft, in the Straits of Dover. It is understood that, in accordance with international law, such passage will be exercised in a continuous and expeditious manner".[416]

Finally, we must point out that this legislative competence of the coastal States of article 37 straits which is regulated by article 42, is applicable, "mutatis mutandi", to the 'archipelagic straits' which are designated as archipelagic sea

[414]See *BOE* No. 81, 4 April 2007, p. 14668.

[415]See http://www.un.org/Depts/los/convention_agreements/convention_declarations.htm.

[416]See http://www.un.org/Depts/los/LEGISLATIONANDTREATIES/regionalist.htm, in order to see the text of the three treaties cited.

lanes by the archipelagic States, with the exception of the designation of sea lanes and the establishment of schemes for the separation of traffic which will be carried out following the stipulations in article 53 of the UNCLOS.

To date, the only archipelagic State which has developed this legislative competence has been Indonesia. In fact, on June 28, 2002, Indonesia adopted State Regulation No. 37 whereby the rights and obligations of foreign vessels and aircraft which exercise the right of passage through the archipelagic sea lanes are regulated and these sea lanes are designated. As a consequence of this routing, as we pointed out in Chap. 4, the straits of *Karimata, Sunda, Makassar, Lombok, Leti* and *Ombai* have been designated as archipelagic sea lanes, therefore, a right of passage similar to transit passage governs these, while, in the other Indonesian straits, the right of innocent passage with the possibility that it might be suspended is applied.[417]

6.2.2.2 The Executive Power of the Coastal States as Regards Transit Passage Through the Straits

The third Section of Part II of the 1982 Convention recognises the right of the coastal State to adopt all the measures required to prevent any passage which is not innocent, to regulate the cases in which it may exercise criminal and civil jurisdiction as regards the vessels which navigate through its territorial sea in innocent passage, and it grants it the right to expel any warship which does not comply with its laws and regulations related to passage through its territorial sea from the strait. The second Section of Part III does not mention this point at all. The only reference appears in paragraph 5 of article 42, but there is no reference to the adoption of measures concerning the infringement of transit passage or its legislation by the coastal State. However, it does refer to the possibility that the coastal State has to demand the international liability of the flag State in the cases of failure of State ships to comply with its laws and regulations; a law which is similar to article 31 analysed above. A right which is a logical consequence of the immunity these types of vessels have under International Law.[418]

The absence of a competence clause similar to that of innocent passage as compared with transit passage, presupposes inhibiting any State territorial control over the vessels intending to cross the strait, and the guarantees of the interests of this State, especially as regards safety, on the good will of the vessels in transit

[417]In conformity with the provisions in article 53 of the 1982 Convention there are three archipelagic sea lanes designated by Indonesia; the last of these contains four variants. State Regulation No. 37 of 2002 and the archipelagic sea lanes established by Indonesia with its corresponding map can be found in the *Law of the Sea Bulletin*, No. 52, 2003, p. 21–41.

[418]However, as regards innocent passage, article 30 legitimates the coastal State to expel these State vessels if they infringed the laws and regulations concerning innocent passage, a right which is not recognised in transit passage.

Does this mean that the coastal State lacks any executive competence as regards transit passage? Evidently no; but it does show that this is much more limited and that, in order to provide grounds for this, it is necessary to carry out a kind of magic as regards the interpretation of the pertinent provisions of Part III.

What happens when a ship or an aircraft crossing a strait in transit passage carries out an activity which the coastal State judges to be incompatible with the rules of 'transit passage'? What can a coastal State do as regards passage which is not 'in transit'? One example is the case of ship or an aircraft which carries out military manoeuvres, exercises with weapons, which might disrupt the communication systems of the coastal State or it simply decides to stop with no force majeure or serious difficulty. Faced with this situation, the Convention does not authorise the coastal State the express power to adopt the measures required to prevent such passage in transit, as authorised by article 25.1 in innocent passage. Thus, we must refer to article 38.3 which sets out the following:

> Any activity which is not an exercise of the right of transit passage through a strait remains subject to the other applicable provisions of this Convention.

Transit passage, which is not adjusted to the parameters set out in article 38, or the conduct patterns of article 39, will be submitted to the pertinent provisions concerning the territorial sea, insofar as this continues to be the legal nature of the waters which include an international strait. In accord with article 25.1, this means that the coastal State is competent as regards the sanctioning of such conduct.

In conclusion, article 38.3 may be invoked as a residual rule to justify an executive intervention of the coastal State as the sovereign State of the territorial, sea. However, T. Treves points out that the coastal State must be prudent on applying article 38.3, as regards the exercise of its sovereign powers in the strait, which must ensure certain proportionality between infringement and its countermeasure.[419]

It is in no doubt that Part III only expressly recognises legislative and not executive authority for the coastal States of straits; however, it is also certain that there is a particular provision – although not within Part III – which recognises the executive competence of the coastal States of international straits although this refers to a very specific material area, which is the "protection and preservation of the marine environment", regulated in Parte XII of the CNDUM. We refer to article 233, which expressly provides that:

> Nothing in sections 5, 6 and 7 affects the legal regime of straits used for international navigation. However, if a foreign ship other than those referred to in section 10 has committed a violation of the laws and regulations referred to in article 42, paragraph 1(a) and (b), causing or threatening major damage to the marine environment of the straits, the States bordering the straits may take appropriate enforcement measures and if so shall respect *mutatis mutandis* the provisions of this section.

[419]Cf. Treves (1985, p. 800).

The objective of this article is to grant powers of application to the coastal States, in the cases involving the infringement of internal laws and regulations dictated by the coastal State concerning the safety of navigation and the regulation of marine traffic, as well as the prevention, reduction and control of pollution by complying with the international regulations applicable regarding the unloading of hydrocarbons, waste petroleum and other harmful substances in the straits. However, this executive power has two important limitations which must be borne in mind. On the one hand, the vessels referred to in Section 10, that is to say, ships which have immunity, such as warships, auxiliary vessels and other ships or aircraft belonging to or used by a State for non-commercial activities are excluded from this competence. On the other hand, this possibility is only restricted to cases in which these infringements have effectively occurred or might cause serious damage to the marine environment of the straits.[420]

Nevertheless, if the conditions required for this provision are complied with, the coastal states could choose to inspect the vessel (article 226) or impose financial penalties for infringing domestic laws and regulations or international rules and norms from among the execution measures stated in the Seventh Section of Part XII, except in the case of an intentional and serious pollution of the territorial sea, in which case, the coastal State can impose sanctions of a criminal nature (article 230).

However, the ambiguous meaning of this article 233 gives rise to many doubts, for example, the appropriate execution measures in specific cases. Besides detaining and inspecting, is it also possible to capture a vessel in transit and adopt criminal jurisdiction measures against the crew?

On this point, there are two possible interpretations. One is to understand that only the execution measures of the Seventh Section can be applied and the other is that the Fifth and Sixth Sections are considered to be equally applicable.

In this regard, in the opinion of V. Carreño Gualde,[421] in order to protect the marine environment, the execution measures stipulated in article 220.2 of the UNCLOS can also be applied, taking into account that these measures may include the physical inspection and immobilisation of the infringing vessel, thus affecting the unhampered exercise of its transit passage. Likewise, R. Riquelme Cortado[422] states that she understands that, besides the execution measures of the Seventh Section, the coastal State could adopt the measures of the Fifth and Sixth Sections, as stipulated in articles 216, 220 and 222, including the retaining and immobilisation of the vessel during its passage through the strait.

In our opinion, if we attend to the literal meaning of the Convention, the execution measures of the Fifth and Sixth Sections are applicable, but only to the extent that they do not hinder, impair or suspend the right to transit passage.

---

[420]The restrictive terms of this article 233 led to a declaration by Spain (the 7th) at the time the Convention was signed, in the sense that this should be interpreted in accordance with the provisions of article 34. However, this declaration is one of those which were not maintained when the UNCLOS was ratified.

[421]See Carreño Gualde (1999, p. 194).

[422]See Riquelme Cortado (1990, pp. 162–166).

However, we believe that the *damage* which these vessels may cause in the waters of the coastal State of the strait must be borne in mind as article 233 permits the adoption of execution measures when non-compliance causes or threatens to cause **serious damage** to the marine environment of a strait. An execution measure which impedes, hinders or hampers the right of transit passage even though it infringes a navigation rule, if it occurs in order to prevent greater damage to the coastal State, it seems to be reasonable and not cost incurring. Clearly, the difficulty involves evaluating the damage to the coastal State and to the user in order to determine which the greater damage is.[423]

In any case, the evaluations of J.A. Yturriaga Barberán are not surprising as, in his opinion, "the special provision on safeguards contained in article 233 is one of the least fortunate of the Convention and implies a denial of the sovereignty of the coastal State over a part of its territorial sea, in contradiction to customary and conventional rules of International Law".[424] Undoubtedly, this provision is another triumph of the strategic considerations as opposed to ecological interests.

In order to escape from the clear rigidity of article 233, the coastal States of the Straits of *Malacca* and *Singapore* reached an understanding with the main users of these straits. The most relevant aspects are included in the Memorandum of April 28, 1982[425] and are the following:

(1) The provisions dictated by the coastal States in accordance with article 42 may include the requirement that vessels in transit through the strait carry a probe under the keel measuring up to 3.5 m. Indonesia, Malaysia and Singapore had already achieved that the IMO recommend this measure five years previously in its Resolution A.375 (X) of 1977.[426]
(2) If this requirement is not complied with, the coastal States can take the appropriate execution measures stipulated in article 233, including the measure to prevent the progress of the vessel, and this cannot be considered to be a refusal, a hindrance, an impairment or a suspension of transit passage.

---

[423]The coastal State can even justify the infringement of the right of transit passage and exclude the illicit aspect of its conduct by alleging the excluding circumstance of the 'state of necessity' stipulated in article 25 of the Project of the ILC concerning the international responsibility of States 2001. Specifically, 'ecological state of necessity' in the sense included in the *Gabcıkovo-Nagymaros Project Case* (cf. ICJ *Reports*, 1997, pp. 51–53), again taking up the opinion of the ILC, for which, a grave danger to the ecological preservation of all or some of the territory of a State included among the situations what could occasion a 'state of necessity', and adds that, "the ecological balance has come to be considered an 'essential interest' of all States" (See *Yearbook of the ILC*, 1980, vol. II, 2nd part, pp. 34, 3, 38 and 14).

[424]See Yturriaga Barberán (1991a, p. 185).

[425]The content of the Memorandum was officially confirmed by Australia, the United States, France, Indonesia, Japan, the United Kingdom and the Federal Republic of Germany. Cf. Doc. A/CONF.62/L.145, of 29 April 1982.

[426]Among other measures, this Resolution recommends that the vessels carry a 3.5 m probe under the keel in order to cross the Malacca and Singapore Straits.

(3) Articles 42 and 233 will not affect the rights nor the obligations of the coastal States as concerns the suitable execution measures regarding vessels which are not in transit passage through these straits.

Italy and France, the coastal States of the *Strait of Bonifacio* created an International Marine Park in the area and wanted to forbid the transit of certain vessels through the Strait in order to prevent and avoid the risk of pollution, a measure which could not be adopted as it was incompatible with the regime of transit passage. What they did do was to adopt internal norms forbidding the traffic of its domestic vessels which transport hydrocarbons, chemical products or other hazardous substances through the straits. This is included in the French Prefecture Resolution No. 1/3, of February 15, 1993, and the Decree of the Italian Merchant Navy Ministry of February 26, 1993[427]

What they did achieve was that the IMO adopted Resolution A.766 (18), of November 4, 1993 on navigation through the *Strait of Bonifacio*, which recommended the governments "to prohibit or at least strongly discourage the transit of the Strait of Bonifacio by laden oil tankers and ships carrying dangerous chemicals or substances in bulk liable to pollute, in the event of a casualty, the sea or the coast, as listed in the annex to MEPC.49 (31) resolution adopted on July 4, 1991, and which are flying the flag of their respective State",[428] although its Preamble reaffirms that freedom of navigation in international straits is an absolute priority.

6.2.3 Obligations of the Coastal States of Straits Concerning Transit Passage

Following the logic of the rights attributed to the coastal State of a strait, the second section of Part III ends with a rule which specifies the obligations of the coastal State with regard to the right of transit passage; obligations which are also compulsory for the archipelagic States as regards 'archipelagic straits' which they might have designated as archipelagic sea lanes. On this point, article 44 stipulates the following:

> States bordering straits shall not hamper transit passage and shall give appropriate publicity to any danger to navigation or overflight within or over the strait of which they have knowledge. There shall be no suspension of transit passage.

Following the meaning of this rule, we can state that the coastal State has both abstention and information obligations.

---

[427]Cf. http://www.un.org/Depts/los/LEGISLATIONANDTREATIES/regionalist.htm.

[428]See the reference in Scovazzi (1995, pp. 150–151).

6.2.3.1 Abstention Obligations

From the provisions in article 44 it is possible to deduce a number of *abstention obligations*, that is to say, things not to do.

The first of these is *not to hinder transit passage*. An obligation which we must place in relation to the construction of fixed links between the two sides of a strait. These constructions which are not very frequent due to their high cost are occasionally constructed in straits used for international navigation,[429] which may entail a 'material' obstacle to the right of transit passage as they may hamper and even prevent navigation. Thus, when they are being built, it is necessary to take this obligation not to hamper transit imposed by the Convention into account, and this involves questions such as their height, leaving sufficiently wide passage for navigation and the possible alterations these types of constructions might cause to the sea and wind currents, which might hinder passage through the strait.

There are fixed links already constructed, such as the one in the *Little Belt*, (one built in 1935 which is 33 m high, and another in 1970 which is 43 m high), in the *Bosphorus* (one built in 1973 and another in 1988, both of which are 64 m high), in the *Great Belt*[430] (inaugurated in August 1997), and in the *Öresund* (opened in July 2000, it combines a tunnel measuring 4,050 m long and a bridge which is 7,845 m long).

The projects for the *Strait of Gibraltar* (where preference has been given to a tunnel and which is now stopped), and the *Dover Strait* (temporally abandoned due to its high cost).

At the present time, planning is being made on the construction of a fixed connection in the *Malacca Strait* (between Indonesia and Malaysia, near Singapore), in the *Johor Strait*, and in the *Corinth Channel*, which was financed by the European Investment Bank.

Undoubtedly, the construction of fixed links in international straits comes within the scope of the sovereignty of the coastal State or States as these are territorial waters, however, this is restricted by the obligation these States have not to hinder the right of passage in transit through these straits, and therefore these constructions must strictly respect this obligation.

In this same context, we must situate the "land reclamation" work carried out by Singapore in the *Johor Strait* and the adjacent zones since 2003. This activity was the subject of an action before the *International Tribunal for the Law of the Sea*

---

[429] For a detailed study of this problem, see the work of Piaskowski-Rafowicz (2004, pp. 319–345).

[430] The construction of the bridge in the *Great Belt* by Denmark was the subject of a claim by Finland submitted to the ICJ on May 17, 1991. On May 23, 1991, Finland requested the adoption of provisional measures by the ICJ, specifically, the stoppage of work. However, this request for provisional measures was rejected by the Court in its Order of July 29, 1991, as it was understood that the work on the construction of a fixed link in the Great Belt did not infringe any right of Finland (cf. *Passage through the Great Belt Case*, ICJ *Reports*, 1991, p. 35). Finland and Denmark finally reached an amicable arrangement which included the payment of 16 million dollars by Denmark to Finland, which meant that Finland ceased its claim before the Court.

brought by Malaysia on July 4, 2003,[431] as it understood that the activity infringed its rights in the zone. On April 26, 2005, Malaysia and Singapore signed a Conciliation Agreement which put an end to the case and constituted the integral and definitive solution to the controversy. Among other questions, Singapore guarantees that, "even after the Palau Tekong land reclamation, the unhampered passage of vessels through Kuala Johor and the port of Calder will not be affected by this land reclamation".[432]

This obligation does not impede transit passage is also stated in paragraph 2 of article 42, when it states that the application of the laws and regulations of the coastal State will not result in denying, hampering or impairing the right of transit passage.

In this regard, as we have mentioned before, it should be remembered that the coastal States of straits where transit passage is in force have not yet made the required changes in their internal legislation in order to adapt it to the new regime, thus, only the general legislation on territorial seas applies. This means that some of these States have provisions in their internal law which, *a priori*, could be included within the material scope of article 42.1(b), that is to say, "prevention, reduction and control of pollution", and these suppose a violation of this obligation "not to deny, hinder or impair the right to transit passage". This is the case of article 5 of Federal Law No. 19 of 1993 of the United Arab Emirates,[433] which establishes the obligation of vessels which transport hazardous or harmful substances to previously notify of their passage. Article 8 of Law No. 45 of 1977 of Yemen[434] imposes this same requisite concerning the passage of these types of vessels. Article 14.2 of Law No. 18 of 1999 of Samoa[435] imposes the requirement that authorisation be granted previously so that nuclear driven vessels and those which transport hazardous substances might cross its territorial waters. This requisite is also demanded by article 9 of the 1993 *Law on Marine Areas* of Iran.[436]

It is evident that the adaptation of domestic legislation in these cases is not only necessary but imperative. Although we must remember that the United Arab Emirates and Iran are not parties to the UNCLOS although Yemen and Samoa are.

In addition, as we pointed out on examining the right of innocent passage, some coastal States of straits in which transit passage is in force also condition the passage of warships through their territorial waters, either by making previous notification obligatory – such as Denmark, Indonesia and India –, or demanding authorisation – such as China, the United Arab Emirates, Iran, St. Vincent and the

[431]Malaysia also requested the adoption of provisional measures by the Court on September 5, 2003, which was decided by the Order of October 8, 2003 (cf. http://www.itlos.org, Case No. 12).

[432]Point 8 of the Conciliation Agreement of April 26, 2005. The text can be seen in the *Law of the Sea Bulletin*, No. 59, 2005, pp. 30–32.

[433]Cf. *Law of the Sea Bulletin*, No. 25, 1994, p. 94.

[434]Cf. http://www.un.org/Depts/los/LEGISLATIONANDTREATIES/regionalist.htm.

[435]Cf. *Ibidem*.

[436]Cf. *Law of the Sea Bulletin*, No. 24, 1993, p. 11.

Grenadines, Sri Lanka and Yemen.[437] The corresponding change in their legislation in order to adapt it to the regime of transit passage through their straits, which prevents these types of measures, is inexcusable.

In this regard, and similar to the right of innocent passage, the coastal State *cannot suspend transit passage* through the strait. Despite this obligation, in September 1988, Indonesia announced the temporary closure of the passage of any vessel through the *Straits of Sunda* and *Lombok* due to the carrying out of air and naval tactical operations. This suspension was protested by the 12 members of the European Community, as well as by Australia, the United States of America and Japan,[438] as it meant a clear infringement of this obligation.

Even more controversial was the adoption of Turkey in 1994 of the "Maritime Traffic Regulations for the Turkish Straits and the Marmara Region", which entered into force in July 1994.[439] This regulation stipulates the possibility to temporarily impede marine traffic in the Turkish straits and in the region of Marmara in certain cases (perforations, salvage operations, accidents, pollution, etc.), and renders the passage of nuclear driven vessels and those which transport hazardous substances subject to the previous authorisation of the Turkish authorities. This polemic situation does not arise due to the infringement of article 44 which does not occur as the straits are beyond the scope of Part III as they are regulated by a long standing convention (article 35 c), but because some of the provisions adopted contradict the terms of the 1936 Montreaux Convention, which has led to several States, such as Greece and the Russian Federation, protesting.

The *prohibition to make distinctions* between foreign vessels in the laws and regulations adopted by the coastal State in relation to the subjects stated in article 42.1 also involves an abstention obligation. This obligation is included in paragraph 2 of article 42, not in article 44, and constitutes an important innovation to the Convention.

6.2.3.2 Obligations Regarding Information and Publicity

In addition, the coastal State must comply with other obligations which entail a *duty of information and publicity*, very similar to what we have analysed as regards innocent passage.

This includes the obligation pointed out in article 44, in the sense that any danger which might threaten navigation or overflight in the strait must be notified of as it becomes known. The determination of the "appropriate way" will depend on the circumstances in each case.

However, due publicity must be given to all its laws and regulations concerning transit passage (article 42.3), and the charts showing the sea lanes and the schemes for the separation of traffic established by the State (article 41.6). The choice of the

---

[437] See above heading 1.1 (C).

[438] Cf. *AJIL*, vol. 83, 1989, p. 559.

[439] Cf. *Law of the Sea Bulletin* No. 27, 1995, pp. 56–72.

publicity method is the exclusive competence of the coastal State, and a suitable means must be used so that the information might reach all the parties involved, that is to say, users of the straits, flag and registration States of the ships and aircraft which cross the strait and the international organisms competent in this regard.

6.2.4 Cooperation Between Coastal States of a Strait and the States Using the Strait

Finally, we must refer to a provision which stipulates international cooperation between coastal States and States using the straits, which has no equivalent in innocent passage. This involves article 43, which sets out the following:

> User States and States bordering a strait should by agreement cooperate:
> (a) in the establishment and maintenance in a strait of necessary navigational and safety aids or other improvements in aid of international navigation; and
> (b) for the prevention, reduction and control of pollution from ships.

The idea contained in this provision is that the compensation for services provided by the coastal State of a strait should be a question to be resolved between the State which uses the strait and the coastal State, since, as we stated above, the regime of Part III does not recognise any right for the coastal State to impose levies for determined services rendered, as occurs in innocent passage.

This provision is based on a proposal of the United Kingdom,[440] which recognises that international interest in navigation through international straits imposes certain restrictions on the rights of the coastal states to these straits, and proposes cooperation between coastal States and the flag States of the vessels and the registration States of the aircraft which use the strait as the most advisable in relation to the navigation aids and the prevention of pollution from the vessels. This article was particularly intended for the *Malacca Strait*.[441]

Although article 43 does not specify how this cooperation must be carried out, it may be direct or through the proper international organisation, which, in this case, would be the IMO, although there are no international instruments which regulate this cooperation.

In this regard, the Foreign Secretaries of Indonesia, Malaysia and Singapore made a Joint Ministerial Declaration on August 2, 2005,[442] whereby they urged the user States of the Straits of Malacca and Singapore to provide assistance to the coastal States in the areas of development of capacity, instruction and the transfer of technology, as well as greater collaboration. Along these same lines, on September 8, 2005, these three States signed the Declaration of Jakarta[443] which, among other

[440]Cf. Doc. A/CONF.62/C.2/L.3, article 3.

[441]Cf. Nandan and Anderson (1989, p. 193).

[442]Cf. *Law of the Sea Bulletin*, no. 59, 2005, pp. 40–41.

[443]Cf. *Ibidem*, pp. 41–45.

questions, agreed to encourage the tasks focused on the application of article 43 of the Convention, to periodically meet with the user States, the shipping sector and other parties concerned in order to examine questions related to safety and protection of the environment in the straits and to facilitate cooperation, as well as to create mechanisms for exchanging information between these. Moreover, the Governments of Indonesia, Malaysia and Singapore requested the IMO to study the possibility to convene complementary meetings so that the coastal States might determine their necessities and the user States in order to study the possibility of convening complementary meetings so that the coastal States might determine their needs and the user States might determine the assistance to attend to these necessities, for example, in the form of information, instruction, training and technical support. The International Maritime Organization, as part of its Protection of Vital Shipping Lanes Initiative, decided to convene a high-level conference to address the security of ships plying the Straits of Malacca and Singapore in collaboration with the three littoral states of Indonesia, Malaysia, and Singapore. Three meetings were held over the period 2005–2007. The outcome was the creation of a framework for cooperation between the users of the straits and the littoral States known as the *Cooperative Mechanism*, a historic breakthrough as Article 43 of the Law of the Sea Convention was implemented for the first time.[444]

In addition, we must point out that this regulation is intentionally drafted in conditional and not in imperative terms as is demonstrated by the use of the expression "they should", instead of "they must", which does impose obligation and is used for this purpose in many other precepts of the Convention. Therefore, we can conclude that the cooperation between user States and coastal States of straits intended by article 43 is a declaration of principles rather than an authentic legal obligation for third party States.

6.3 Regulation of the Safety of Navigation and Sea Traffic: Sea Lanes, Schemes for the Separation Sea Traffic and Other Safety Measures Adopted in the Straits Where Passage in Transit Is in Force

According to the provisions in article 41, the coastal States of straits can designate lanes and establish schemes for navigation in transit through the straits, but not with freedom as in innocent passage but when the following conditions arise:

1. That it is necessary for the safe passage of the vessels.
2. That the sea lanes and schemes are adjusted to generally accepted international regulations. We must understand these to be Regulation V/8 of SOLAS, Rules 1

[444] The decisions made at the three meetings, the analysis of the responses of the various stakeholders during the meetings, and the explications of the reasons for the success of the Cooperative Mechanism with interesting conclusions by proposing ways in which, can be consulted in the works of Ho (2009, pp. 233–247), and Beckman (2009, pp. 233–259).

(d) and 10 of COLREG 1972,[445] and the General Provisions on Ships' Routeing of the IMO contained in Resolution A.572 (14), as amended by Resolution MSC.71 (69). In addition, we must take Resolution A.858 (20) *regarding the Procedure for the adoption and amendment of traffic separation schemes* into consideration. Also of interest are Resolution A.857 (20) which establishes the *Guidelines for Vessel Traffic Services*, Resolution A.851 (20) which includes the General Principles for Ship Reporting Systems and Ship Reporting Requirements, including *Guidelines for Reporting incidents involving Dangerous Goods, Harmful Substances and/or Marine Pollutants.*

3. That the proposals have been submitted for adoption to the competent international organisation, which is the IMO. Of course, the IMO cannot adopt a decision in this regard without the agreement of the coastal States of the strait, but these States cannot establish more sea lanes or schemes than those adopted by the Organisation. In the case of straits with two or more coastal States, these will cooperate in order to formulate proposals in consultation with the IMO.

Therefore, following article 41, the process for the establishment and for the replacement of sea lanes and schemes for traffic safety is as follows: the proposal of the coastal State, the adoption of the Organisation in accord with this, and the establishment or substitution by an internal state act. In the case of two or more coastal States, these will cooperate in order to present a joint proposal to the IMO.

There is also the paradoxical case put forward by S.C. Truver[446] that the coastal States might propose a determined scheme and that the IMO refuses to adopt this, and this cannot be adopted unilaterally. This would generate an *impasse* difficult to solve, a dead end, which would mean not adopting any measures to guarantee the safety of navigation in the strait, nor to order traffic, with the consequent harm to the interests of international navigation in safe routes. This is aggravated by the fact that submarines can navigate submerged in these straits, which increases the risk.

Only **three international straits** where transit passage is in force have schemes for the separation of traffic or other navigation safety systems, adopted by the IMO. However, among these are the five most important straits for international navigation, namely, *Dover*, *Gibraltar*, *Bab-el Mandeb*, *Hormuz* and *Malacca-Singapore*.

– Little Minch.

As we have pointed out in the previous chapter, the IMO has jointly adopted the prohibition of passage by oil tankers weighing more than 10,000 tons through the two *Minches*, as a safety measure, unless there is force majeure (SN/Circ.159).[447]

– North Channel.

As a safety measure, the IMO has prohibited the passage of tankers through the *Rathlin Sound passage* (SN/Circ.159).

---

[445] See the text in Chap. 5.

[446] See Truver (1980, pp. 203–204).

[447] The measure affects the two *Minches – North* and *Little –* although the first comes under the regime of innocent passage (article 38.1) and the second is ruled by transit passage (article 37).

– ***The Hole***.

In the region of the Shetland Islands, the IMO has established two precaution areas, one to the north and another to the west and south. This second area is the one which affects this strait. In order to avoid the risk of oil pollution and severe damage to the environment, vessels of more than 5,000 GT carrying oil or other hazardous cargoes in bulk should avoid the area bounded by lines connecting the following geographical positions (SN/Circ.159 and SN/Circ.167):

(10) 60°02′.50 N, 1°10′.20 W; Heli Ness
(11) 59°59′.87 N, 1°09′.37 N; Perie Bard Lt
(12) 59°41′.0 N, 1°12′.0 W
(13) 59°42′.7 N, 1°26′.0 W
(14) 60°02′.0 N, 2°48′.0 W
(15) 60°15′.0 N, 2°48′.0 W
(16) 60°42′.5 N, 2°09′.0 W
(17) 60°42′.5 N, 1°22′.0 W
(18) 60°40′.0 N, 1°17′.0 W
(19) 60°37′.3 N, 1°17′.9 W

Thence along the West coastline to position (10).

– ***Dover Strait***.

On January 1, 1967 the first traffic separation scheme adopted by the IMO came into force (SN/Circ.35) and this has undergone several modifications, the last of which were in 2006 in relation to post F3 located at northeast edge of the Strait (COLREG.2/Circ.57), and in 2007 (COLREG.2/Circ.59).The first, 2006, introduces the following amendments:

- The existing separation line passing through the F3 station is deleted.
- The geographical positions of the boundary for the new "Precautory Area" around the F3 Station Buoy are as follows (co-ordinates are based on WGS 84 Datum):

  (1) 51°26′.01 N, 002°02′.67 E
  (2) 51°25′.31 N, 002°03′.81 E
  (3) 51°23′.23 N, 001°58′.69 E
  (4) 51°22′.76 N, 001°59′.59 E

- The position of the F3 Station Buoy and the area surrounding it in *IMO Ships' Routeing 7th Edition 1999, Part D, I/4* remain unchanged.
- Recommended direction of traffic flow arrows is inserted in accordance with convention for ships crossing the Precautory Area around the F3 Station, passing the buoy and leaving it on their own port side as follows:

  1. To the northeast of the F3 Station Buoy indicating a north-westerly traffic flow
  2. To the southwest of the F3 Station Buoy indicating a south-easterly traffic flow

The Governments of Belgium, France and United Kingdom have proposed to the 53rd session of the Sub-Committee on Safety of Navigation of IMO a new change in the existing traffic separation scheme (NAV. 53/3/15, 53/3/17/ and 53/3/18). The Maritime Safety Committee adopted on October 31, 2007 this new amendment for the traffic separation scheme in the Strait of Dover and Adjacent Waters (COLREG.2/Circ.59), with the following provisions:

(a) A separation zone is bounded by lines connecting the following geographical positions:

(1) 51°25′.31 N, 002°04′.03 E
(2) 51°26′.77 N, 002°01′.48 E
(3) 51°31′.07 N, 002°07′.90 E
(4) 51°29′.84 N, 002°10′.62 E

(b) A separation line connects the following geographical positions:

(5) 51°26′.97 N, 002°16′.95 E
(6) 51°22′.83 N, 002°12′.29 E

(c) A separation zone is bounded by lines connecting the following geographical positions:

(7) 51°22′.03 N, 001°58′.39 E
(8) 51°22′.49 N, 001°57′.61 E
(9) 51°16′.53 N, 001°52′.29 E

(d) A precautionary area with recommended directions of traffic flow is established connecting the geographical positions (1), (2), (8) and (7) above.

(e) A separation line connects the following geographical positions:

(10) 51°16′.53 N 001°52′.29 E
(11) 51°06′.13 N 001°38′.10 E

(f) A separation zone is bounded by lines connecting the following geographical positions:

(12) 51°05′.77 N 001°38′.65 E
(13) 51°06′.49 N 001°37′.55 E
(14) 50°57′.59 N 001°23′.00 E
(15) 50°51′.14 N 001°17′.20 E
(16) 50°33′.37 N 000°36′.50 E
(17) 50°26′.91 N 000°01′.09 W
(18) 50°22′.12 N 000°00′.91 E
(19) 50°32′.71 N 000°57′.73 E
(20) 50°42′.87 N 001°18′.30 E
(21) 50°56′.87 N 001°24′.03 E

(g) A traffic lane for south-westbound traffic is established between the separation zones/lines described in paragraphs (a), (c), (e) and (f) above and the following separation line/zone:

a separation line connection the following geographical positions:

(22) 51°33′.66 N 002°02′.17 E
(23) 51°27′.35 N 001°52′.76 E
(24) 51°14′.13 N 001°43′.99 E
(25) 51°06′.93 N 001°30′.90 E
(26) 50°52′.29 N 001°02′.65 E

a separation zone bounded by lines connecting the following geographical positions:

(27) 50°52′.47 N 001°02′.45 E
(28) 50°39′.37 N 000°32′.50 E
(29) 50°34′.64 N 000°04′.29 W
(30) 50°32′.71 N 000°03′.49 W
(31) 50°38′.91 N 000°32′.70 E
(32) 50°52′.09 N 001°02′.85 E

(h) A traffic lane for north-eastbound traffic is established between the separation zones/lines described in paragraphs (a), (c), (e) and (f) above and the following separation line/zone:
a separation zone is bounded by lines connecting the following geographical positions:

(33) 50°16′.34 N 000°03′.31 E
(34) 50°14′.49 N 000°04′.11 E
(35) 50°26′.37 N 001°00′.20 E
(36) 50°39′.29 N 001°22′.63 E
(37) 50°39′.69 N 001°22′.20 E
(38) 50°26′.94 N 000°59′.90 E

a separation line connects the following geographical positions:

(39) 50°39′.49 N 001°22′.40 E
(40) 50°44′.54 N 001°26′.90 E
(41) 50°53′.64 N 001°30′.70 E
(42) 51°04′.34 N 001°45′.89 E

a separation zone is bounded by lines connecting the following geographical positions:

(43) 51°04′.34 N 001°45′.89 E
(44) 51°06′.44 N 001°48′.89 E
(45) 51°11′.23 N 002°04′.09 E
(46) 51°09′.84 N 002°03′.12 E

an uncharted line representing the junction of the scheme with the adjacent scheme at West Hinder and joining the following geographical positions:

(47) 51°11′.23 N 002°04′.09 E

(6) 51°22′.83 N 002°12′.29 E

A separation zone is established within this lane as described in (i) below.

(i) A separation zone is bounded by the lines connecting the following geographical positions:

(48) 51°18′.43 N 002°04′.69 E
(49) 51°16′.03 N 002°04′.19 E
(50) 51°13′.71 N 002°00′.99 E
(51) 51°09′.35 N 001°47′.10 E
(52) 51°09′.75 N 001°45′.61 E
(53) 51°12′.35 N 001°51′.03 E
(54) 51°15′.05 N 001°54′.40 E

(j) A deep-water route forming part of the north-eastbound traffic lane between the separation zone described in (i) above and the separation zone/line described in paragraphs (c) and (e) above has been established between a line connecting the following geographical positions:

(i) 51°09′.75 N – 001°45′.61 E
(ii) 51°10′.26 N – 001°43′.74 E;

and

(iii) 51°22′.03 N – 002°58′.39 E
(iv) 51°18′.43 N – 002°04′.69 E

Inshore traffic zones

The area between the outer boundary of the traffic separation scheme and the English coast which

lies between a line:

(v) 51°08′.42 N 001°22′.24 E
(vi) 51°02′.53 N 001°22′.24 E

and a line between:

(vii) 50°34′.64 N 000°04′.29 W
(viii) 50°49′.60 N 000°16′.86 W

is designated as an inshore traffic zone.

The area between the outer boundary of the traffic separation scheme and the French coast which lies between:

(ix) 50°53′.64 N 001°30′.70 E
(x) 50°52′.10 N 001°34′.96 E

and a line between:

(xi) 50°30′.09 N 001°06′.66 E
(xii) 50°30′.09 N 001°34′.59 E

is designated as an inshore traffic zone.

Warnings

1. A deep-water route forming part of the north-eastbound traffic lane is established to the north-west of the Sandettie Bank, and masters considering the use of this route should take into account the proximity of traffic using the south-westbound lane.
2. The main traffic lane for north-eastbound traffic lies to the south-east of the Sandettie Bank and shall be followed by all such ships as can safely navigate therein having regard to their draught.
3. In the area of the deep-water route east of the separation line, ships are recommended to avoid overtaking.

It is important that ships passing through the Dover Strait listen to the appropriate VHF broadcasts y the Channel Navigation Information Service which provide information concerning traffic, navigation and visibility conditions in the Strait.

On the other hand, the Maritime Security Committee of the IMO has adopted on January 8, 1999 a "Mandatory Ship Reporting System" for the ships of 300 gross tonnage (SN/Circ.205), and has adopted on October 8, 2007 some amendments to the existing mandatory reporting system in Dover Strait/Pas de Calais, into force at 00.00 hours on May 1, 2008 (MSC.251 (83)). Likewise as the Dover Strait is a part of the *English Channel* also is applicable the Resolution A.486 (XII), November 19, 1981, regarding "Recommendation on the use of adequately qualified deep-sea pilots in the North Sea, English Channel and Skagerrak" (SN/Circ.167, June 7, 1994).

– Strait of Gibraltar.

The last amendment to the existing the separation scheme in the Strait of Gibraltar was adopted by the IMO on December 11, 2006 (COLREG.2/Circ.58) with the following description:

Traffic Separation Scheme

(a) A separation zone, half a mile wide, is centred upon the following geographical positions:

(1) 35°59′.01 N, 005°25′.68 W
(2) 35°58′.36 N, 005°28′.19 W

(b) A separation zone, half a mile wide, is centred upon the following geographical positions:

(3) 35°57′.08 N, 005°33′.08 W
(4) 35°56′.21 N, 005°36′.48 W
(5) 35°56′.21 N, 005°44′.98 W

(c) A traffic lane for westbounded traffic is established between the separation zone described in paragraph (a) and a line connecting the following geographical positions:

(7) 36°01′.21 N, 005°25′.68 W
(8) 36°00′.35 N, 005°28′.98 W

(d) A traffic lane for westbounded traffic is established between the separation zone described in paragraph (b) and a line connecting the following geographical positions:

(9) 35°59′.07 N, 005°33′.87 W
(10) 35°58′.41 N, 005°36′.48 W
(11) 35°58′.41 N, 005°44′.98 W

(e) A traffic lane for eastbounded traffic is established between the separation zone described in paragraph (b) and a line connecting the following geographical positions:

(12) 35°52′.51 N, 005°44′.98 W
(13) 35°53′.81 N, 005°36′.48 W
(14) 35°54′.97 N, 005°32′.25 W

(f) A traffic lane for eastbounded traffic is established between the separation zone described in paragraph (a) and a line connecting the following geographical positions:

(15) 35°56′.35 N, 005°27′.40 W
(16) 35°56′.84 N, 005°25′.68 W

(g) A precautionary area is established on the eastern side of the Gibraltar TSS by the lines connecting the following geographical positions:

(6) 36°02′.80 N, 005°19′.68 W
(7) 36°01′.21 N, 005°25′.68 W
(16) 35°56′.84 N, 005°25′.68 W
(17) 35°58′.78 N, 005°18′.55 W

(h) A precautionary area with recommended directions of traffic flow is established off the Moroccan port of Tanger-Med in the Gibraltar TSS formed by the lines connecting the following geographical positions:

(8) 36°00′.35 N, 005°28′.98 W
(9) 35°59′.07 N, 005°33′.87 W
(14) 35°54′.97 N, 005°32′.25 W
(15) 35°56′.35 N, 005°27′.40 W

Inshore traffic zones

North Zone: The area between the northern boundary of the scheme formed by the continuing line that links points 7, 8, 9, 10 and 11 and the Spanish coast, and lying between the following limits is designated as an inshore traffic zone:

(a) Eastern limit: that part of the meridian 005°25′.68 W (23) between the northern boundary of the westbound traffic lane (latitude 36°01′.21 N, corresponding to point (7) on the attached chartlet) and the Spanish coast.

(b) Western limit: that part of the meridian 005°44′.98 W (22) between the northern boundary of the westbound traffic lane (latitude 35°58′.41 N, corresponding to point (11) on the attached chartlet) and the Spanish coast.

South Zone: The existing southern inshore traffic zone is divided into two inshore traffic zones to east and west, with a free navigational area between them, located between the southern limit of the TSS and the coast of Morocco; these are bounded by eight geographical positions.

South-eastern zone:

(18) 35°45′.45 N, 005°25′.68 W
(16) 35°56′.84 N, 005°25′.68 W
(15) 35°56′.35 N, 005°27′.40 W
(19) 35°54′.88 N, 005°27′.40 W

South-western zone:

(20) 35°51′.33 N, 005°32′.25 W
(14) 35°54′.97 N, 005°32′.25 W
(12) 35°52′.51 N, 005°44′.98 W
(21) 35°49′.09 N, 005°44′.98 W

The IMO has also established a "Notification System for Vessels" in the area of the traffic separation scheme, which arose from a proposal made by Spain in cooperation with Morocco, on August 12, 1996 (NAV.42/23, p. 5.4). This system affects all the vessels with a length of more than 50 m, all the vessels which transport hazardous substances as per the definition given by resolution MSC.43 (64), the vessels are over 50 m long are hooked to or pushed by another vessel, any category of vessel shorter than 50 m but which uses the traffic lane or the separation zone in order to work, and any type of vessel shorter than 50 m which is using the traffic lane or the separation zone in an emergency in order to avoid an imminent danger (SN/Circ.186).

– Strait of Bonifacio.

As regards this strait, the IMO has adopted several navigation measures different of the traffic separation schemes. Specifically IMO has established two routes for navigation of vessels which are over 20 m long:

Northern limit: a line joining the geographical positions:

A. 41°22′.55 N, 009°22′.38 E
F. 41°18′.00 N, 009°15′.25 E
E. 41°19′.18 N, 009°06′.51 E

Southern limit: a line joining the geographical positions:

B. 41°21′.58 N, 009°23′.30 E
C. 41°16′.75 N, 009°15′.75 E
D. 41°16′.75 N, 009°06′.18 E

and the establishment of a number of precaution areas in these two routes which affect all ships (SN/Circ.198).

Eastern precautionary area: A circular sector pointed on geographical position M: 41°22′.05 N 009°22′.85 E with a radius of 5 (five) nautical miles limited by lines joining geographical positions A (see above) and G: 41°26′.90 N 009°24′.50 E and joining geographical positions B (see above) and H: 41°19′.31 N 009°28′.40 E.

Western precautionary area: A circular sector pointed on geographical position N: 41°17′.96 N 009°06′.33 E with a radius of 5 (five) nautical miles limited by lines joining geographical positions E (see above) and L: 41°21′.37 N 009°01′.47 E and joining geographical positions D (see above) and I: 41°13′.57 N 009°03′.15 E.

It has also established a notification system for ships of 300 gross tonnage and over (SN/Circ.201, 26 May 1998).

– Strait of Bab-el-Mandeb.

The latest amendment of the traffic scheme in force for this strait was adopted by the IMO in 2003 (COLREG.2/Circ.52). It is composed of two traffic sea lanes and a zone for the separation of traffic between them. The direction of navigation will be:

- A southbound traffic lane, 155° (T) as far as the turning line off Mayyun Island, thence 120° (T) to the eastern limit of the existing scheme.
- A northbound traffic lane, 300° (T) as far as the turning line off Mayyun Island, thence 335° (T) to the northern limit of the existing scheme.

Traffic Separation Scheme

(a) A separation zone bounded by a line connecting the following geographical positions:

(1) 13°13′.07 N, 043°02′.87 E
(2) 12°36′.82 N, 043°20′.22 E
(3) 12°32′.53 N, 043°27′.79 E
(4) 12°33′.37 N, 043°28′.30 E
(5) 12°37′.50 N, 043°21′.00 E
(6) 13°13′.83 N, 043°03′.60 E

(b) A traffic lane for southbound traffic between the separation zone and a line connecting the following geographical positions:

(7) 13°11′.94 N, 043°01′.72 E
(8) 12°35′.78 N, 043°18′.98 E
(9) 12°31′.25 N, 043°27′.04 E

(c) A traffic lane for northbound traffic between the separation zone and a line connecting the following geographical positions:

(10) 13°15′.00 N, 043°04′.70 E
(11) 12°38′.50 N, 043°22′.21 E
(12) 12°34′.69 N, 043°29′.03 E

– Strait of Hormuz.

This strait has a traffic separation scheme 6 miles wide, composed of two traffic lanes which are 2 miles wide each one – one to way in and other to way out –, an intermediate traffic separation zone 2 miles wide and an associated inshore zone (COLREG.2/Circ.11).

– ***Straits of Malacca*** **and** ***Singapore.***

The IMO has adopted several measures to guarantee the security of the navigation in the Straits of Malacca and Singapore.

There is in force a traffic separation scheme in which has established a Precautionary area in the Strait of Singapore south of Tanjong Stapa bounded by a line connecting the following geographical positions (COLREG.2/Circ.13):

(i) 01°14′03″N, 104°03′35″E
(ii) 01°16′36″N, 104°03′19″E
(iii) 01°18′38″N, 104°15′00″E
(iv) 01°15′24″N, 104°15′00″E

The deep water routes in the Singapore Strait are amended as follows:

(i) 01°03′36″N, 103°38′57″E
(ii) 01°05′54″N, 103°43′23″E
(iii) 01°08′36″N, 103°45′26″E
(iv) 01°10′15″N, 103°47′54″E
(v) 01°09′57″N, 103°48′17″E
(vi) 01°08′54″N, 103°46′49″E
(vii) 01°04′57″N, 103°42′52″E
(viii) 01°02′58″N, 103°39′06″E
(ix) 01°10′15″N, 103°47′54″E
(x) 01°11′18″N, 103°50′30″E
(xi) 01°12′07″N, 103°52′12″E
(xii) 01°11′48″N, 103°52′38″E
(xiii) 01°09′57″N, 103°48′17″E

The Maritime Safety Committee, as its 78h session, adopted, in accordance with the provisions of resolution A.858 (20), an amendment to the traffic separation scheme in the Singapore Strait for the establishment of an anchorage area in the existing separation zone for emergency anchoring of vessels including damaged vessels for repairs prior to entry into a shipyard or similar matters. On June 28, 2004, the IMO received a copy of the text of a Decree of the Directorate General of the Sea Communications No. PP 72/8/1-03 dated November 20, 2003 on "Standard operating procedures for management and operation of the Nipah transit anchorage area on the waters of Nipah" from the Government of the Republic of Indonesia (SN/Circ.238, COLREG.2/Circ.54), which is settled on the following geographical positions:

(a) To exit:

(85) 01°10′.35 N, 103°34′.90 E
(86) 01°10′.35 N, 103°39′.85 E
(87) 01°07′.50 N, 103°43′.72 E
(88) 01°08′.60 N, 103°45′.43 E

(89) 01°05′.90 N, 103°43′.38 E
(90) 01°03′.60 N, 103°38′.98 E
(91) 01°07′.06 N, 103°32′.96 E

(b) To entry:

(85a) 01°09′.40 N, 103°36′.60 E
(86a) 01°09′.10 N, 103°38′.60 E
(89a) 01°05′.50 N, 103°40′.80 E
(90a) 01°04′.50 N, 103°38′.90 E
(91a) 01°06′.80 N, 103°35′.00 E

On the other hand, it has established a naval traffic information service for the Strait of Singapore (SN/Circ.143), a group of 11 Rules to regulate navigation in both straits through the traffic separation scheme (SN/Circ.198), and a notification system for vessels. This last measure affects vessels over 300 gross tonnage, those which are 50 m or more in length, those which exceed 300 gross tonnage or 50 m when combined with other vessels or are pushed by others, the vessels which transport hazardous substances, the passengers of such vessels, and any category of vessel weighing less than 300 gross tonnage or measuring less than 50 m which might use the sea lane or the separation zone in the event of an emergency in order to avoid immediate danger (SN/CIRC.201).

– East Lamma Channel.

To date, Hong Kong Marine Department and MSA China have proposed a traffic separation scheme in the Zhujiang area in 2003. This proposal includes the recommendation to establish also another scheme for the Dangan Channel and the Lantau Channel, to the south west of the Soko Islands (PLVAC Paper No.6/2003). The IMO has not yet answered.

– Torres Strait.

The IMO has adopted routeing measures other than traffic separation schemes. It has established a double direction route in the Great Northeast Channel of the Torres Strait in the following geographical positions (SN/Circ.234):

(a) Northern limit:

1. 10°29′.70 S, 142°22′.63 E
2. 10°29′.14 S, 142°25′.76 E
3. 10°27′.80 S, 142°28′.45 E
4. 10°26′.40 S, 142°31′.30 E
5. 10°21′.90 S, 142°41′.50 E
6. 10°19′.37 S, 142°47′.97 E
7. 10°18′.14 S, 142°50′.82 E
8. 10°13′.38 S, 142°54′.96 E
9. 10°00′.50 S, 143°03′.42 E
10. 09°47′.73 S, 143°10′.40 E
11. 09°25′.80 S, 143°31′.07 E
12. 09°12′.47 S, 143°51′.34 E

(b) Southern limit:

13. 10°30′.45 S, 142°24′.02 E
14. 10°28′.38 S, 142°28′.66 E
15. 10°27′.38 S, 142°31′.85 E
16. 10°22′.85 S, 142°41′.95 E
17. 10°19′.80 S, 142°48′.23 E
18. 10°17′.63 S, 142°53′.29 E
19. 10°09′.78 S, 143°05′.55 E
20. 09°53′.97 S, 143°15′.61 E
21. 09°46′.02 S, 143°18′.48 E
22. 09°37′.96 S, 143°21′.97 E
23. 09°27′.60 S, 143°32′.15 E
24. 09°13′.95 S, 143°52′.62 E

(c) Centre polygon:

25. 10°16′.10 S, 142°53′.82 E
26. 10°13′.79 S, 142°55′.85 E
27. 10°01′.05 S, 143°04′.20 E
28. 09°48′.10 S, 143°11′.30 E
29. 09°41′.04 S, 143°18′.87 E
30. 09°45′.72 S, 143°17′.51 E
31. 09°53′.84 S, 143°14′.50 E
32. 10°09′.15 S, 143°04′.70 E

Therefore, IMO has established a mandatory ship reporting system (SN/Circ.235 and MSC/78/26/Add. 2, May 17, 2004), which includes information on sea traffic, navigation aid, and information on marine safety. The Maritime Safety Committee has established the mandatory pilotage in the inner route of the Great Barrier Reef (MSC.79).

– ***Bass Strait***.

The IMO has also established a traffic separation scheme which has been modified on several occasions. In the COLREG.2/Circ.24 (June 24, 1985) it has been established the following scheme:

(a) A separation zone, half a mile wide, is centred upon the following geographical positions:

(1) 38°41′.05 S, 148°20′.02 E
(2) 38°44′.05 S, 148°14′.09 E
(3) 38°46′.03 S, 148°09′.00 E

(b) A traffic lane for westbound traffic between the separation zone and a line connecting the following geographical positions:

(4) 38°38′.05 S, 148°17′.05 E
(5) 38°41′.00 S, 148°13′.02 E
(6) 38°46′.03 S, 148°07′.03 E

(c) A traffic lane for eastbound traffic between the separation zone and a line connecting the following geographical positions:

(7) 38°49′.08 S, 148°10′.08 E
(8) 38°48′.00 S, 148°16′.07 E
(9) 38°44′.06 S, 148°23′.00 E

The Government of Australia informed the IMO that it will make a temporary change to one of the TSS "In the Bass Strait" on the south east coast of Australia at 00:00 hours UTC on March 1, 2002. The adjustment is to allow for the drilling of petroleum exploration wells (COLREG.2/Circ.50). Revised geographical coordinates of the salient points of the TSS are as follows:

(a) A separation zone, 6.0 miles wide centred upon the following geographical positions:

1. 38°43′.40 S, 148°21′.66 E
2. 38°46′.33 S, 148°16′.37 E
3. 38°48′.32 S, 148°10′.17 E

(b) A traffic lane for westbound vessels between the separation zone and a line connecting the following geographical positions:

4. 38°38′.50 S, 148°17′.50 E
5. 38°41′.00 S, 148°13′.20 E
6. 38°42′.80 S, 148°07′.30 E

(c) A traffic lane for eastbound vessels between the separation zone and a line connecting the following geographical positions:

7. 38°53′.92 S, 148°13′.03 E
8. 38°51′.72 S, 148°20′.09 E
9. 38°48′.40 S, 148°26′.17 E

– Old Bahama Channel.

This strait has a traffic separation scheme on the north coast of the Island of Cuba proposed by the Cuban Ministry of Transport, and a reporting system for ships (SN/Circ.141/Rev.1, February 15, 1993).

6.4 The Right of Transit Passage: A Conventional Norm Generally Accepted as Law?

The impact the 1982 Convention with regard to the norms of the Law of the Sea has been impressive due, among other things, to the substantial progressive development it has provided about the matter. It is obvious that the right of transit passage through the straits used for international navigation is part of this.

In this regard, one of the most relevant questions which has been of concern to international law experts since the adoption of the UNCLOS on December 10, 1982 arose. This was the determination whether the right of transit passage has become a customary norm of General International Law.[448] J.N. Moore made an ominous prediction when he stated that, "with or without a new convention… the UNCLOS strait regime seems destined to serve as a powerful model for the development of a new customary law of straits transit".[449]

This is a question which was of decisive importance throughout the 14 years in which the Convention was not yet in force, and, after the entry of the Convention into force, on November 16, 1994, it continued to be important, taking into account the fact that a large number of coastal States of article 37 straits were clearly reticent to becoming a party to the Convention. However, ratifications such as that of Spain in 1997, Malaysia in 1996, and Greece in 1995, the only three 'survivors' of the "Group of Strait States" and the strongest opponents to transit passage while defending innocent passage through all international straits at all costs, together with the ratifications of other States which were not excessively enthusiastic about the new regime in the course of the III Conference, and are also coastal States of straits where transit passage must be in force, such as Morocco (2007), Denmark (2004), Canada (2003) and Ukraine (1999), have considerably reduced the importance of this question.

The problem of the customary nature of the right of transit passage is clearly of interest as regards the States which are not parties to the UNCLOS. We must take into account that there are five coastal States used for international navigation which have not ratified nor have adhered to the 1982 Convention: the United Arab Emirates, Iran, the U.S.A., East Timor, and Venezuela. Neither is Turkey a party, but the straits where this is a coastal State (the *Bosphorus* and the *Dardanelles*) are separated from Part III, as these are regulated by a long standing convention (*1936 Montreux Convention*). We should point out four of these States: the United Arab Emirates, Iran, the U.S.A. and Venezuela, as these are coastal States which comply with the parameters of article 37 and, consequently, transit passage must be applied in these if it is concluded that the customary norm exists in these and these States are bound by this. Although the figure is reduced to three on excluding the United States, which undoubtedly recognises the existence and validity of the right to transit passage through international straits.[450] This position remained evident in the course of the III Conference (as we saw in the second chapter), as was subsequently seen when it was expressly included in its domestic

---

[448]Among the abundant bibliography on this point, we can cite, Bernhardt (1987, pp. 249–331); Caminos (1987, pp. 178–235); Jia (1998, pp. 168–208); Treves (1986, pp. 247–259); Treves (1985, pp. 805–808); or Treves (1990, pp. 9–302).

[449]See Moore (1980, p. 121).

[450]In addition, it seems that the U.S.A. will ratify the UNCLOS and will become a party to it. This can be deduced by the Declaration made by President G.W. Bush, on May 15, 2007, in which he stated that "I urge the Senate to act favourably on U.S. accession to the United Nations Convention on the Law of the Sea during this session of Congress" (see http://www.whitehouse.gov/news/releases/2007/05/print/20070515-2.html).

law (Proclamation of 1988[451]). In any case, the reasons for this hooligan claim for transit passage made by the United States are well known to be grounded on its own interest rather than on the interest of the international community.

In the event that the response is negative, that is to say, it is not possible to definitively state that there is a norm which is generally accepted as law and which recognises the right to transit passage in international straits which join an area of the high seas or an Exclusive Economic Zone, or the conclusion is reached that these States are in no case bound by it as they have manifested an attitude which is clearly contrary to the norm from the beginning, the result would be that, in the part of the territorial sea of the straits where these three States are coastal States, the regime of navigation would be the right of innocent passage, a customary norm generally accepted as law. The straits in question are those of *Hormuz*, the *Aruba-Paraguana P*assage, the Serpent's *Mouth* and the *Dragon's Mouth*, all of which are vitally important for international navigation, and the *Strait of Hormuz* is a fundamental strategic route.

As regards this point, H. Caminos and M.R. Molitor pointed out that the legal effect of the 'package deal' is that "it cannot affect those provisions of the Convention which are carried over directly from the 1958 Conventions and which reflected customary law prior to UNCLOS III. Such provisions, irrespective of their place within the 1982 package, continue to be exercisable by, and binding upon, third States. However, where established customary rules appearing in the 1982 Convention have been changed, it must not be presumed that such modifications have acquired similar status".[452] In our opinion, in the case of transit passage this does not involve any of the matters where the Convention is a reflection of international customary law, and the States which are not parties to the Convention cannot object.

As the ICJ stated in the *North Sea Continental Shelf Case*,[453] it is not sufficient that there is a broad participation in a Convention for it to be automatically transformed into a customary law. Therefore the legal relations which arise from instruments concluded during bilateral negotiations are not of a customary nature *per se*, but are simply contractual. The true test for the existence of a customary norm of International Law is that of State practice, understood in its most general acceptance as the group of actions carried out individually or collectively by the States, and which constitute the most important indication in the determination of the *opinio iuris* and, therefore, of the existence of the norm itself.

In this regard, therefore, the practice of the coastal States of international straits which are not parties to the UNCLOS is the only way to identify whether the right of transit passage is a norm of General International Law. In order to generate this

[451]Cf. *ILM*, vol. 8, 1989, p. 284. For an examination of the position of the U.S.A. in this regard, see Wainwright (1986, pp. 361–414).

[452]See Caminos and Molitor (1985, p. 887).

[453]Cf. ICJ *Reports* 1969, p. 42.

effect, this practice must be uniform, constant and generalised, moreover, it must be carried out in such a way that the legal conviction that this practice is obligatory is demonstrated, that is to say, that there is an *opinio iuris*.

In this regard, the pertinent practice fundamentally refers to the domestic legislation of the States concerning navigation through the international straits, especially as regards coastal States of the straits of article 37, the unilateral declarations which have been made in this regard, and the international treaties concerning this. All of these points have been dealt with in this work.

Related to this summit, we can see that only three States have incorporated the reference to transit passage in their straits to their internal legislation,[454] the United States, the United Kingdom and Spain. The reasons of the first two are more than evident, therefore, they are not a decisive factor in this case. As regards Spain, we must understand that the incorporation of an express reference to the regime of transit passage must be understood as the logical consequence of its ratification of the Convention no less than 10 years later, thus it can only be deduced that this regime was accepted and not conventionally so. Another two States, Cuba and Djibouti, have included the reference to the freedom of navigation and overflight in straits in their legislation, but not expressly referring to transit passage but to the principles of International Law, thus their position is not clear. This is even less clear if we take into account that Djibouti demands previous notification from nuclear driven vessels and those which transport hazardous substances in order to pass through their territorial sea, including *Bab-el-Mandeb*, and submarines must pass on the surface which is incompatible with the transit passage regime which must be in force in this strait.

No other State has incorporated the transit passage regime into its internal legislation, not even mentioning his. This did not occur before nor after they became parties to the UNCLOS, therefore, do their laws which recognise innocent passage in their straits continue to be in force. Some of these even have provisions, as we have seen, which infringe the obligation not to hinder transit passage. This is the case, among others, of the *Federal Law* of 1993 of the United Arab Emirates, and the *Law on Marine Zones* of 1993 of Iran, States which are not parties to the UNCLOS, which include the right of innocent passage as the only regime applicable in its territorial sea. A position which is also taken up in the 1956 *Territorial Law of the Sea* of Venezuela.

As regards the declarations in the *Passage through the Great Belt Case*, Denmark declared that the transit passage regime regulated in Part III of the 1982 Convention cannot be applied to the case as customary law, substituting the regime of the 1958 Convention.[455] Such a declaration clearly contrasts with the declaration of Finland regarding the same matter, which affirms that there is a strong tendency

---

[454]We must remember that the internal legislation concerning marine areas can be consulted at http://www.un.org/Depts/los/LEGISLATIONANDTREATIES/regionalist.htm.

[455]Cf. Counter Report of Denmark, p. 744 (http://www.icj-cij.org).

in international practice towards the establishment of customary law in transit passage.[456]

As we have pointed out, when Spain signed and ratified the Convention, it made an interpretative declaration of the adverb "normally", tending to restrict the overflight of State aircraft above straits.[457]

However, on signing the UNCLOS on December 10, 1982, Iran expressly stated the following:

> Notwithstanding the intended character of the Convention being one of general application and of law making nature, certain of its provisions are merely product of quid pro quo which do not necessarily support to codify the existing customs or established usage (practice) regarded as having an obligatory character. Therefore, it seems natural and in harmony with article 34 of the 1969 Vienna Convention on the Law of Treaties, that only states parties to the Law of the Sea Convention shall be entitled to benefit from the contractual rights created therein.
>
> The above considerations pertain specifically (but no exclusively) to the following:
>
> – The right of transit passage through straits used for international navigation (Part III, Section 2, article 38).[458]

More recently, Morocco ratified the UNCLOS, on May 31, 2007, and declared, as we stated above, that "the laws and regulations relating to maritime areas in force in Morocco shall remain applicable without prejudice to the provisions of the United *Nations Convention on the Law of the Sea*".[459] This entails that its Law of 1973 remains in force, whose article 3 establishes the regime of innocent passage through the *Strait of Gibraltar*.

Finally, we have previously mentioned some treaties concerning maritime limitation between the coastal States of international straits. Thus, the Treaty made by Australia and Papua New Guinea concerning the sovereignty and maritime frontiers of the two countries, including the *Torres Strait*, made on December 18, 1978, recognises the freedom of navigation and overflight in the Torres Strait in its article 7, which is the equivalent of a recognition of the regime of transit passage. As regards this same point, the Joint Declaration of the status of the *Dover Strait* made by France and the United Kingdom, attached to their Agreement of November 2, 1988 *on the limitation of their maritime border*, states that both governments recognise the right of transit passage through this strait[460]

However, Venezuela has made two sea frontier delimitation agreements with its neighbouring States and these clarify its non-recognition of transit passage. Thus,

---

[456]Cf. Report of Finland, p. 394 (http://www.icj-cij.org).

[457]Cf. http://untreaty.un.org.

[458]See http://untreaty.un.org.

[459]See http://www.un.org/Depts/los/convention_agreements/convention_declarations.htm.

[460]The texts of the Treaties made by Australia and Papua New Guinea, and the United Kingdom and France can be seen at http://www.un.org/Depts/los/LEGISLATIONANDTREATIES/regionalist.htm.

article 4 of the Treaty signed by Venezuela and the Netherlands on March 31, 1978, recognises the right of transit passage between the Island of Aruba and mainland Venezuela, that is to say, in the *Aruba-Paraguana Passage,* but only for the respective ships and aircraft, as article 4.3 excludes State aircraft and warships of third party States from such passage.[461] Article VI of the Treaty of April 18, 1990, made by Venezuela and Trinidad and Tobago, provides that innocent passage be applied in the Gulf of Paria;[462] a provision which affects the *Serpent's Mouth* and the *Dragon's Mouth.*

Two important *conclusions* can be drawn from preceding practice:

• **One**, practice concerning the application of the right of transit passage, as well as its recognition as a customary norm of a general nature, is neither constant nor uniform. On the one hand, there are only five States which have shown a clear and evident attitude to accepting transit passage as a generally accepted right: the United States, the United Kingdom, France, Australia and Papua New Guinea.

On the other hand, if we understand that the axiom "practice generally accepted as law" implies that, if the right of transit passage is a custom, "it should also be granted by such States as if they were fulfilling a duty of customary law",[463] we find that this meaning of duty is not complied with in many States. Specifically it is not complied with by Canada, Chile, Colombia, Cuba, Denmark, Djibouti, Egypt, Greece, Indonesia, Iran, Japan, Republic of Korea, Malaysia, Morocco, Oman, Panama, the Philippines, Spain, Trinidad and Tobago, Venezuela and Yemen.[464]

In fact, it cannot be stated that the tendency towards customary implementation of navigation within the regime of transit passage has been as uniform as it should have, not even in the case of the Great Sea Powers, as is shown in the attitude of the ex-Soviet Union, which is clearly ambivalent as it claims transit passage for all the straits except for those under its jurisdiction.

Therefore, we can conclude that the customary statute of the right to transit passage is still uncertain, and it cannot be stated that it has achieved the nature of customary norm in General International Law.

Thus, as pointed out by D.L. Larson, "the conventional rights and privileges of transit and archipelagic sea lanes passage can only be legally obtained through state accession or ratification".[465]

---

[461] Cf. http://www.un.org/Depts/los/LEGISLATIONANDTREATIES/regionalist.htm.

[462] Cf. *Law of the Sea Bulletin*, No. 19, 1991, p. 25.

[463] Cf. Jia (1998, p. 202).

[464] Cf. *Ibidem.*

[465] The author adds that, "If the United States remains outside of the legal regime established by the 1982 Convention, it cannot claim or obtain the rights and privileges thereof without assuming the 'correlative duties' of the deep seabed mining". See Larson (1987, p. 428).

• **Two**: Moreover, if we focus on practice followed by the three coastal straits of the article 37 straits which are not parties to the UNCLOS, we can see that Iran and Venezuela have again stated their evident positions as 'persistent objectors',[466] thus, there is no doubt regarding the disregard doctrine of a 'hypothetical right of customary transit passage'. The same conclusion could also be applied as regards the United Arab Emirates since, although these have not declared their opposition to the regime of transit passage when signing the UNCLOS in 1982, their internal legislation of 1993 concerning passage through their territorial seas imposes requirements which are a clear attack on passage transit.

This means that the Strait of *Hormuz*, the *Aruba-Paraguana Passage, the Dragon's Mouth and the Serpent's Mouth* are regulated by innocent passage as a generally accepted customary law which obliges their coastal States.

[466]Venezuela did not even sign the 1982 Convection.

Conclusions

I – The regulation of passage through the straits is a question which has been of concern both for doctrine and for the States, from the very origins of the Law of the Sea. However, except for some straits which had their own conventional regimes, Turkish and Danish Straits and the Strait of Magellan, it is only well into the twentieth century that a general legal regime concerning straits is specified.

The International Court of Justice with its decision on the *Corfu Channel Case* of April 9, 1949 and the article 16.4 of the *Geneva Convention on Territorial Sea and Contiguous Zones* of 29 April 1958, consolidated the application of the right to innocent passage in all the straits used for international navigation as a customary norm of general international law within the general legal framework of the territorial sea.

II – The progressive increase of the width of the territorial sea up to 12 miles supposed the parallel increase in the importance of the legal regulation of passage through straits as this means that more than a hundred straits, including those which were strategically important, those where there was freedom of navigation and overflight since there is a high seas corridor, are now subject to the coastal State or States as they involve waters which are territorial seas. This situation, together with the appearance of other factors such as the increase in the number and nature of ships and aircraft which use the straits, the growing importance of the straits for the transport of oil, the deployment of military naval and air power, and the consequent risk for the contamination and safety of the coastal State have led to the split between coastal States, defenders of innocent passage through these straits, and the States which use the straits, especially, the naval powers, which claim freedom of navigation and overflight through these straits, and the need for revision which became more acute at the end of the seventies.

This confrontation and the need for revision led to the III Conference of the United Nations on the Law of the Sea, which ended with the stipulations in Part III of the *United Nations Convention on the Law of the Sea 10 December 1982*. This supposes an autonomous legal regulation for the straits, apart from the territorial sea, which sets out a duality of navigation regimes depending on the type of strait used for the navigation in question: the right of innocent passage for secondary straits, the right of transit passage for main straits.

III – Despite the revision carried out by the 1982 Convention, a question which is as fundamental as the concept of "straits used for international navigation" was maintained in the same terms as in the 1958 Convention, that is to say, as undefined.

Taking the doctrine of the States and of the International Court of Justice as a reference, the concept of straits used for international navigation necessarily involves the confluence of three components: geographical, legal and functional. Only those straits which are affected by the three components come under the legal regime of Part III, which supposes a total of 101 international straits.

IV – The functional component – namely that the strait is used for international navigation – is the most crucial of the three, but the most difficult to determine and which entails greater discrepancy of opinions. Once an interpretation has been made, we understand that this component is complied with insofar as the strait in question is being used for international navigation, regardless of the level and the type of use; the only requirements are that the use is current and navigation is marine and on the surface.

The coastal State of a strait is not obliged to carry out work or provide assistance to make the strait navigable. However, if it carries out such work or assistance and as a consequence of these the strait is used for international navigation, since thence it is obliged to continue to do so by application of 'estoppel'.

V – There are eight categories of straits differentiated by the 1982 Convention, which add up to a total of 229 straits, only three of these categories are within the scope of Part III, these are the following that. Straits formed by an island of a coastal State and its mainland territory, when there is a free navigation route on the other side of the island (article 38.1), their number amounts to 34. Straits located between a part of the high seas or an exclusive economic zone and the territorial sea (article 45.1b), there are only seven straits of this group. Straits which join high seas or exclusive economic zones (article 37), there are 60 main straits. The first two are regulated by the right of innocent passage and the last by transit passage.

The remaining categories remain apart from Part III, these are the following. Straits which include interior waters which have always been interior (article 35 a), these are 21 and are regulated by Part II. Straits regulated by long standing conventions (article 35 c), there are seven and are regulated by their specific convention. Straits regulated by an agreement compatible with the 1982 Convention (article 311.2), there is one. Straits where there is a high sea route or which cross an exclusive economic zone (article 36), there are 53 straits of this group and these are regulated by Part VII or Part V, depending on the case. And straits which involve archipelagic waters in an archipelagic State amount to 47 and are regulated by the provisions in Part IV.

VI – The right of innocent passage has become an exceptional rule in the regulation of international navigation through straits, and is limited to two categories which are those with least strategic importance, thus, they are termed "secondary straits". On this point, Part III does not regulate the functioning of this right of passage which is proper to the territorial seas, but restricts itself to a reference to what is stipulated in Part II, and specifies the prohibition o its suspension, in terms identical to the conventional regime of 1958.

Questions which were already controversial in 1958, and whose importance has grown, such as the passage of warships or ships with special characteristics, remain with no clear and adequate regulation. This means that coastal States of straits of this category such as Albania, Estonia, Finland, Sweden and Montenegro demand previous notification of the passage of warships through diplomatic channels so that these may enter their territorial seas including straits. Estonia also demands previous authorisation for nuclear driven vessels and Italy has established obligatory piloting in the Strait of Messina for vessels which transport hazardous substances.

VII – The innovative right of transit passage consecrates the freedom of navigation for warships, submarines and other vessels navigating submerged, as well as the freedom of overflight for all aircraft, including military aircraft, in such terms that this is different from the high seas only by the fact that this navigation and overflight must only take place rapidly and expeditiously. However, there is no sufficiently detailed catalogue of the activities these ships and aircraft are forbidden to carry out during their transit passage.

This "deviation" from the regime of innocent passage supposes the establishment of an unjustified additional easement for the coastal State of a strait, a safety risk and an increase in the probability of accidents and the pollution of these waters.

VIII – As regards the overflight of straits, the Convention introduces minimum safety controls which not only suppose a threat for the safety of the State and the coastal towns of the straits, but also a very serious limitation to their sovereignty. Moreover, these clearly contradict with the safety and control measures imposed by the International Civil Aviation Organisation and the 1944 Chicago Convention. A circumstance which raises an added problem concerning the application of successive treaties related to the same matter as there is not a total coincidence of States parties to the two Conventions.

IX – The right of transit passage arbitrarily restricts the legislative competence of the coastal State, and limits this to the adoption of laws and regulations as regards some aspects of this passage, and even, some of these involve limitations. Thus, as regards the safety of navigation and the regulation of sea traffic, the coastal State proposes but the International Maritime Organisation adopts the measures. In relation to the preservation of the marine environment, it is limited to implementing the applicable regulations concerning the unloading of certain harmful substances.

Among the omissions of the regulating competence of the coastal State of a strait, attention should be drawn to the absence of the recognition of legislative competence concerning air navigation, as well as the omission of any reference to its right to establish air corridors and devices for the separation of air traffic.

X – Except for the exceptional case of the United States, the United Kingdom and, recently, Spain, none of the coastal States of article 37 straits have made the pertinent changes to their internal legislation in order to adapt it to the transit passage regime. Some, such as Morocco, have expressly stated that that they are maintaining their internal legislation previous to the 1982 Convention. Thus, the laws in force are the general laws of territorial seas which include the right of innocent passage as the regime applicable to international straits.

This scenario entails paradoxical situations such as those caused by coastal States of straits where transit passage should be in force and these are parties to the 1982 Convention, but they maintain internal legislation which requires the previous authorisation or notification regarding warships, which is the case of Yemen, India, China, Indonesia and St. Vincent, or nuclear driven vessels or those transporting contaminating substances, as is the case in Yemen, Djibouti and Samoa.

XI – The executive competence of the coastal states of straits is reduced even more as it is not expressly stipulated that it can adopt measures to prevent passage which 'is not in transit', and only recognises a minimum power of execution in the case of ships which fail to comply with the norms concerning the regulation of sea traffic or the unloading of harmful substances which might cause or threaten to cause serious damage to the marine environment in the strait. It is necessary to carry out a conjuring trick with the interpretation of article 38.3 of the Convention in order to provide a base for the logical sanctioning capacity of the coastal State.

XII – The right of transit passage is not of a customary nature. An examination of the practice of States in this matter leads us to conclude the absence of uniformity in this regard, together with the lack of legal conviction.

Furthermore, the position of the coastal States of article 37 straits, which are not parties to the 1982 Convention, is in clear opposition to the regime of transit passage and a firm affirmation of the right of innocent passage. This is the position of Iran, the United Arab Emirates and Venezuela, and this situation affects the Strait of Hormuz, the Aruba-Paraguana Passage, the Dragon's Mouth and the Serpent's Mouth.

XIII – Finally, we can conclude by pointing out that, unlike other 'innovations' to the Law of the Sea made by the 1982 Convention where priority was given to the sovereignty of the coastal States over the freedom of the seas, as well as the exclusive economic zone and archipelagic waters, in the straits used for international navigation the order of primacy is inverted and strategic interests have prevailed over the sovereignty of the coastal States. This also entails an inversion regarding the 1958 Geneva Convention concerning straits, where sovereignty was dominant over freedom.

Thus, the contemporary Law of the Sea has 'internationalised' virtually all the straits in the world, minimised the control of the coastal State as regards sea traffic through these straits. Any attempt by the coastal States of the straits to claim unconditional sovereign control in the straits and over sea traffic using these straits is absolutely unreal in the era of globalising capitalism and during the 'reign' of the only world super power.

Catalogue of Straits

Europe

- PENTLAD FIRTH, between the Orkneands Islands and Scotland, United Kingdom (article 38.1)
- THE HOLE, between the Britannic Islands of Orkneands and Shetland (article 37)
- NORTH MINCH, between the Outer Hebrides and the mainland of United Kingdom (article 38.1)
- LITTLE MINCH, between the Out Hebrides and the English Island of Skande (article 37)
- NORTH CHANNEL, between Scotland and North Ireland (article 37)
- THE SOLENT, between the Island of Wight and the mainland of United Kingdom (article 38.1)
- ST. GEORGE'S CHANNEL, between United Kingdom and Ireland (article 36)
- DOVER or PAS DE CALAIS, between United Kingdom and France (article 37)
- SKAGERRAK, between Denmark and Norway (article 36)
- KATTEGAT, between Denmark and Sweden (article 36)
- BORNHOLMSGATTET or STRAIT OF BORNHOLM, between Denmark and Sweden (article 36)
- ØRESUND or SOUND o SUNDET, between Denmark and Sweden (article 35 c)
- LITTLE BELT, in Denmark (article 35 c)
- GREAT BELT, in Denmark (article 35 c)
- KADET, between Germany and Denmark (article 36)
- KALMARSUND, between the Island of Öland and the mainland of Sweden (article 38.1)
- ÅALAND, between Swede and Finland (article 35 c)
- ENTRANCE TO GULF OF FINLAND, between Estonia and Finland (article 45.1 b)
- IRBEN, between the Island of Saaremaa and the mainland of Latvia (article 45.1 b)
- HARI or HURK, between the Latvian Islands of Hiiumaa and Vormsi (article 45.1 b)
- SUUR, between the Island of Saaremaa and the mainland of Latvia (article 45.1 b)
- GIBRALTAR, between Spain and Morocco (article 37)
- FREU OF MENORCA o CHANNEL OF MENORCA, between the Spanish Islands of Mallorca and Menorca (article 37)
- STRAIT OF BONIFACIO or BOUCHE OF BONIFACIO, between Corsica (France) and Sardinia (Italy) (article 37)
- ELBA STRAIT or PIOMBINO CHANNEL, between the Island of Elba and the mainland of Italy (article 38.1)
- STRAIT OF SICILY, in Italy (article 36)
- CORSICA CHANNEL, between France and Italy (article 36)
- MALTA CHANNEL, between Italy and Malta (article 36)

- MESSINA, between the Island of Sicily and the mainland of Italy (article 38.1)
- OTRANTO, between Albania and Italy (article 36)
- BRACKI KANAL, in Croatia (article 35 a)
- HVARSKI KANAL, in Croatia (article 35 a)
- KORCULANSKI KANAL, in Croatia (article 35 a)
- LASTOVSKI KANAL, in Croatia (article 35 a)
- MIJETSKI KANAL, between Otok Mljet and the Peninsula of Peljesac in Croatia (article 38.1)
- NIKOLA, between the Island of Sveti Nikola and Budva in Montenegro (article 38.1)
- CORFU, between Albania and Greece and the Greek Island of Corfu (article 38.1)
- CORINTH CHANNEL, between Peloponnese and the southeast of Greece (article 38.1)
- KARPATHOS, between the Greek Islands of Karppathos and Rodas (article 37)
- KASOS, in Greece (article 36)
- EURIPUS, between the Island of Euboæa and the mainland of Greece (article 38.1)
- KITHERA CHANNEL or DHIÉKPLOUS KITHIRON, between the Greek Islands of Kithera and Antikithera (article 37)
- KERCHENSKLAND PROLIV o STRAIT OF KERCH, between Russia and the Peninsula Ukrainian of Crimea (article 37)
- BOSPHORUS, in Turkey (article 35 c)
- DARDANELLES, in Turkey (article 35 c)

Africa and Middle East

- TIRAN, between Egypt and Saudi Arabia (article 45.1 b and article 311.2)
- BAB-EL-MANDEB, between Yemen, Djibouti and Somalia (article 37)
- PEMBA CHANNEL, between Kenya and Tanzania and the Island of Pemba (article 38.1)
- ZANZIBAR CHANNEL, between the Island of Zanzibar and the mainland of Tanzania (article 38.1)
- MOZAMBIQUE CHANNEL, between Mozambique and Madagascar (article 36)
- HORMUZ, between Iran – north –, and Oman and the Arab Emirates – south – (article 37)

Asia and South Pacific

- KAMCHATSKIAND PROLIV, in Russia (article 36)
- PROLIV LITKE, between the Island of Karaginskand and the Russian Peninsula of Kamchatka (article 38.1)

- PERVANDAND KURIL'SKIAND PROLIV, between the Russian Peninsula of Kamchatka and the Kuriles (article 37)
- CHETVERTANDAND KURIL'SKIAND PROLIV, in Russia (article 36)
- PROLIV KRUZENSHTERNA, in Russia (article 36)
- PROLIV FRIZA, in the Kuriles (article 37)
- PROLIV ANDEKATERINAND, in the Kuriles (article 37)
- SHIKOTAN SUIDO, in the Kuriles (article 37)
- TARAKU SUIDO, in the Kuriles (article 37)
- NEMURO-KAIKANDO, between the Island Kuril of Yekaterinand and Japan (article 37)
- PROLIV NEVEL'SKOGO and TATARSKIAND PROLIV, between the Island of Sajalin and the Asiatic mainland of Russia (article 38.1)
- STRAIT OF SOANDA o LA PÉROUSE, between Russia and Japan (article 36)
- CHANNEL OF KII, in Japan (article 35 a)
- RISHIRI-SUIDO, between the Island of Rishiri and Hokkaido, in Japan (article 38.1)
- TSUGARU-KAIKANDO, in Japan (article 36)
- SADO-KAIKANDO, between the Island of Randotsu and Honshu, in Japan (article 38.1)
- OSUMI-KAIKANDO or STRAIT OF VAN DIEMAN, in Japan (article 36)
- TANEGA-SHIMA-KAIKANDO, between the Japanese Islands of Tanega-shima and Yaku-shima (article 37)
- ANDAKUSHIMA-KAIKANDO, in Japan (article 36)
- TOKARA-KAIKANDO, between the Japanese Islands of Osumi and Tokara Gunto in the Randukandu (article 37)
- STRAIT OF EAST KOREA or TSUSHIMA KAIKANDO, in Japan (article 36)
- WEST KOREA or STRAIT OF WESTERN CHOSEN, between Korea and the Japanese Island of Tsushima (article 36)
- CHEJU HAEHANDOP, between the Island of Cheju and southwest mainland of Korea (article 38.1)
- TAIWAN HAIXIA, between Taiwan and China (article 36)
- P'ENG-HU SHUI-TAO or CHANNEL OF PESCADORES, between the Island of Pescadores and Taiwan (article 38.1)
- BOHAI HAIXIA or STRAIT OF PO HAI, in China (article 35 a)
- LAMMA or LEMA CHANNEL, between Hong-Kong and the Chinese Island of Lema (article 37)
- HAINAN or QUIONGZHOU, between the Island of Hainan and the mainland in China (article 38.1)
- PALK, between India and Sri Lanka (article 37)
- MACPHERSON, between the Islands of South Andaman and Rutland in the archipelago of Andaman and Nicobar, India (article 37)
- BASHI CHANNEL, between Taiwan and Philippines (article 36)
- BALABAC, between Malaysia and Philippines (article 36)
- MALACCA, between Malaysia and Indonesia (article 37)
- SINGAPUR, between Malaysia and Singapore (article 37)

- BALITANG CHANNEL, in Philippines (part IV)
- BABUANDAN CHANNEL, in Philippines (part IV)
- POLILLO, in Philippines (part IV)
- SAN BERNARDINO, in Philippines (part IV)
- MAQUEDA CHANNEL, in Philippines (part IV)
- VERDE ISLAND PASSAGE, in Philippines (part IV)
- MINDORO, in Philippines (part IV)
- SURIGAO, in Philippines (part IV)
- BASILAN PASSAGE, in Philippines (part IV)
- SAN JUANICO, in Philippines (part IV)
- BOHOL o CEBU, in Philippines (part IV)
- TANON, in Philippines (part IV)
- SIBUTU PASSAGE, in Philippines (part IV)
- LOMBOK, in Indonesia (part IV)
- MAKASSAR, in Indonesia (part IV)
- SERASAN, in Indonesia (part IV)
- BERHALA, in Indonesia (part IV)
- BANGKA, in Indonesia (part IV)
- GASPAR, in Indonesia (part IV)
- SUNDA, in Indonesia (part IV)
- GELASA, in Indonesia (part IV)
- MENTAWI, in Indonesia (part IV)
- KARIMATA, in Indonesia (part IV)
- BALI, in Indonesia (part IV)
- BADUNG, in Indonesia (part IV)
- ALAS, in Indonesia (part IV)
- SAPE, in Indonesia (part IV)
- GREHUND, in Indonesia (part IV)
- KOTI, in Indonesia (part IV)
- API, in Indonesia (part IV)
- TIMPAUS, in Indonesia (part IV)
- MANIPA, in Indonesia (part IV)
- ROTI, in Indonesia (part IV)
- LETI, in Indonesia (part IV)
- OMBAI, in Indonesia (part IV)
- WETAR, between Indonesia and Timor Leste (article 45.1 b)
- TORRES, between Australia and Papua New Guinea (article 37)
- BASS STRAIT, in Australia (article 37)
- BANKS STRAIT, in Australia (article 37)
- ISUMRUD, in Papua New Guinea (part IV)
- VITIAZ, in Papua New Guinea (part IV)
- ST. GEORGE CHANNEL, in Papua New Guinea (part IV)
- BOUGAINVILLE, in Salomon Islands (part IV)
- MANNING, in Salomon Islands (part IV)
- INDISPENSABLE, in Salomon Islands (part IV)

- VATU-I-RA CHANNEL, in Fiji (part IV)
- RANGITOTO CHANNEL, in New Zealand (article 35 a)
- TAMAKI, in New Zealand (article 35 a)
- COLVILLE CHANNEL, between the Island of Great Barrier and the Peninsula of Coromandel, New Zealand (article 38.1)
- COOK, between North Island and South Island in New Zealand (article 37)
- FOVEAUX or FLOVEAUX, between the Stewart and South Island, in New Zealand (article 38.1)
- APOLIMA, between the Islands of Savati and Upolu, in Samoa (article 37)

North America-Pacific

- BERING, between Russia and USA (article 37)
- ETOLIN, between the Island of Nuvivak and Alaska, in USA (article 38.1)
- SHELIKOF, between the Island of Kodiak and Alaska, in USA (article 38.1)
- AGATTU, in the Aleutians Islands, USA (article 37)
- AMCHITKA PASSAGE, in the Aleutians, USA (article 36)
- TANAGA PASSAGE, in the Aleutians, USA (article 37)
- UNIMAK PASSAGE, in the Aleutians, USA (article 37)
- SAMAIGA PASSAGE, in the Aleutians, USA (article 37)
- AMUKTA PASSAGE, in the Aleutians, USA (article 36)
- SEGUAN PASSAGE, in the Aleutians, USA (article 37)
- ADAK, in the Aleutians, USA (article 37)
- ICAND, between the Island of Chicago and Alaska, in USA (article 38.1)
- KAULAKAHI CHANNEL, in Hawaii, USA (article 37)
- KAUAI CHANNEL, in Hawaii, USA (article 36)
- KAIWI CHANNEL, in Hawaii, USA (article 37)
- PAILOLO CHANNEL, in Hawaii, USA (article 37)
- ALENUIHAHA CHANNEL, in Hawaii, USA (article 36)
- ALALAKEIKI CHANNEL, in Hawaii, USA (article 37)
- KEALAIKAHIKI CHANNEL, in Hawaii, USA (article 37)
- AUAN CHANNEL, in Hawaii, USA (article 37)
- KALOHI CHANNEL, in Hawaii, USA (article 37)
- CHATHAM, in Canada (article 35 a)
- DIXON ENTRANCE, between Canada and USA (article 36)
- HECATE, between the Island of Queen Charlotte and the west mainland of Canada (article 38.1)
- QUEEN CHARLOTTE AND GEORGIA, between the Island of Vancouver and west mainland of Canada (article 38.1)
- JUAN DE FUCA, between Canada and USA (article 45.1 b)
- PUGET SOUND, in USA (article 35 a)
- GOLDEN GATE, in USA (article 35 a)

- CHANNEL OF SANTA BARBARA, between the Channel Islands and the mainland of California (article 38.1)

Artique

- McCLURE, in Canada (article 36)
- VICTORIA, in Canada (article 36)
- JAMES ROSS, in Canada (article 36)
- BARROW, between the Islands of Prince Leopold and Melville, in Canada (article 37)
- LANCASTER SOUND, in Canada (article 36)
- PRINCE REGENT INLET, in Canada (article 36)
- VISCOUNT MELVILLE SOUND, in Canada (article 36)
- FOXE CHANNEL, in Canada (article 36)
- HUDSON, in Canada (article 36)
- ROBESON CHANNEL, between Canada and Denmark (article 37)
- NARES, between Canada and Denmark (article 37)
- DAVIS, between Canada and Denmark(article 36)
- STRAIT OF DENMARK or GREENLAND, between Denmark and Iceland (article 36)
- BRITISH CHANNEL, STRAIT OF AUSTRIAN and STRAIT OF MAKHAM, in the Russian Archipelago of Franz Josef (article 37)
- ORLOVSKAIA, in Russia (article 35 a)
- GORLO, in Russia (article 35 a)
- VOSTOCHNAIA, in Russia (article 35 a)
- ZAPADNAIA, in Russia (article 35 a)
- ANZARSKAIA, in Russia (article 35 a)
- ZHIZHGINSKAIA, in Russia (article 35 a)
- KIL'DIN, between the Island of Kil'din and the mainland of Russia (article 38.1)
- SHOKAL'SKII, in Russia (article 37)
- PROLIV MATOCHKIN SHAR, in Russia (article 37)
- STRAIT OF KARA, in Russia (article 37)
- PROLIV VIL'KITSKOGO or STRAIT OF VIL'KITSKII, in Russia (article 36)
- PROLIV DMITRIANDA LAPTEV, in Russia (article 36)
- PROLIV LONGA, in Russia (article 36)

North America-Atlantic

- BELLE ISLE, between the Island of Newfounland and the Canadian Peninsula of Labrador (article 38.1)

- JACQUES CARTIER PASSAGE, between the Island of Anticosti and the mainland of Canada in Quebec (article 38.1)
- CABOT, in Canada (article 36)
- HONGUEDO o PASSAGE OF GASPE, in Canada (article 36)
- NORTHUMBERLAND, between the Island of Prince Edward and the Peninsula of Nova Scotia, Canada (article 38.1)
- HEAD HARBOUR PASSAGE, between the Peninsula of Nova Scotia and the Gulf of Maine, Canada (article 45.1 b)
- ARTHUR KILL, in USA (article 35 a)
- KILL VAN KULL, in USA (article 35 a)
- THE NARROWS, in USA (article 35 a)

Caribbean

- FLORIDA, between USA and Cuba (article 36)
- CHANNEL OF ANDUCATAN, between Cuba and Mexico (article 36)
- OLD BAHAMA CHANNEL, between Cuba and Bahamas (article 37)
- NORTHWEST PROVIDENCE CHANNEL, in Bahamas (part IV)
- NORTHEAST PROVIDENCE CHANNEL, in Bahamas (part IV)
- PASSAGE OF CROOKED, in Bahamas (part IV)
- PASSAGE OF MAANDAGUANA, in Bahamas (part IV)
- PASSAGE OF TURKS ISLAND, between the Britannic Islands of Turks and Caicos (article 37)
- WINDWARD PASSAGEAGE, between Cuba and Haiti (article 36)
- CHANNEL OF JAMAICA, between Jamaica and Haiti (article 36)
- MONA PASSAGE, between Dominic Republic and Puerto Rico (article 36)
- ANEGADA PASSAGE, in the Britannic Virgin Islands (article 36)
- GUADELOUPE PASSAGE, between Guadeloupe – France – and Montserrat – United Kingdom – (article 37)
- DOMINICA PASSAGE, between Guadeloupe – France – and Dominica (article 37)
- MARTINIQUE PASSAGEAGE, between Dominica and Martinique – France – (article 37)
- ST. LUCIA CHANNEL, between Martinique – France – and St. Lucia (article 37)
- SAINT VICENT PASSAGE, between St. Lucia and Saint Vincent (article 37)
- GALLEONS PASSAGE, in Trinidad and Tobago (part IV)
- SERPENT'S MOUTH, between Trinidad and Tobago and Venezuela (article 37)
- DRAGON'S MOUTH, between Trinidad and Tobago and Venezuela (article 37)
- ARUBA-PARAGUANA PASSAGE, between the Venezuelan Peninsula of Paraguana and Aruba – Dutch Antilles – (article 37)

South America

- MAGELLAN, between Chile and Argentina (article 35 c)
- CHANNEL OF BEAGLE, in Chile (article 35 a)
- STRAIT OF LE MAIRE, between Tierra de Fuego and the Island of Los Estados, in Argentina (article 38.1)
- BRANSFIELD, between the South Shetland Islands and the mainland of Antarctic Peninsula (article 36)
- DRAKE PASSAGE, between the Cap Horn and the South Shetland Islands –between South America and the Antarctica – (article 36)

Bibliography

Alexander L (1986) Navigational restrictions within the new LOS context: geographical implications for the United States. Offshore Consultants, Rhode Island

Alexander LM (1987) Exceptions to the transit passage regime: straits with routes of similar convenience. Ocean Develop Int Law 18:479–491

Alexandersson G (1982) The Baltic Straits. Sijthoff and Noordhoff, The Hague

Amato d' A (1983) An alternative to the law of the sea convention. Am J Int Law 77

Amerasinghe CF (1974) The problem of archipelagos in the international law of the sea. Int Comp Law Q 1974:539–575

Amin SH (1981) The regime of international straits: legal implications for the Strait of Hormuz. J Marit Law Comm 12

Anand RP (ed) (1980) Law of the sea. Caracas and beyond. Martinus Nijhoff, The Hague

Anand RP (1996) Navigation through territorial sea and straits revisited. Ind J Int Law 4:13–38

Anand RP (2004) Navigation through territorial sea and straits. In: Anand RP (ed) Studies in international law and history: an Asian perspective. Martinus Nijhoff, Leiden, pp 197–224

Andrew (1978) Archipelagos and the law of the sea: island straits states of island-studded sea space. Marine Policy 1978(1):46–64

Bardonnet D, Virally M (1983) La nouveau droit international de la mer. Pedone, Paris

Bateman S (2009) Compulsory pilotage in the torres strait: overcoming unacceptable risks to a sensitive environment. Ocean Dev Int Law 40.2:184–203

Baxter (1964) The law of international waterways with particular regard to international canals. Harvard University Press, Cambridge

Beckman R (2009) The establishment of a cooperative mechanism for the straits of Malacca and Singapore under Article 43 of the United Nations Convention on the Law of the Sea. In: Chircop A, McDorman TL, Rolston SJ (eds) The future of the ocean regime building: essays in tribute to Douglas M. Johnston. Martinus Nijhoff, Leiden, pp 233–259

Bernhardt R (1987) Custom and treaty in the law of the sea. R des C 1987(VII):249–332

Bonassies P (1973) Les routes imposées de l'OMCI. In: SFDI, Actualités du droit de la mer. Ed. A. Pedone, Paris, pp 48–61

Bordunov VD (1988) The right of transit passage under the 1982 convention. Marine Policy 1988:219–230

Bou Franch V (1994) La navegación por el mar territorial, incluidos los estrechos internacionales y las aguas archipelágicas. Madrid

Brown ED (1973) Passage through the territorial sea, straits used for international navigation and archipelagos. London

Brown ED (1994) The international law of the sea, vol I, introductory manual. Dartmouth

Brubaker RD (2001) Straits in the Russian Arctic. Ocean Develop Int Law 3:263–287

Brüel E (1947) International Straits. A treatise on international law. Sweet and Maxwell, London

Burke WT (1977) Submerged passage through straits: interpretation of the proposed law of the sea treaty text. Washington Law Rev 1977:193 et seq

Burke KM, De Leo DA (1983) Innocent passage and transit passage in the United Nations Convention on the Law of the Sea. Yale J World Public Order 1983:389 et seq

Burmester H (1982) The Torres Strait Treaty: ocean boundary delimitation by agreement. Am J Int Law 76

Butler WE (1978) Northeast Arctic passage. Sijthoff and Noordhoff, Alphen aan den Rijn Sijthoff

Caminos H (1987) The legal regime of straits in the 1982 United Nations Convention on the Law of the Sea. R. des C. 205(V):9–246

Caminos H (1989) Los estrechos exceptuados del régimen jurídico de los estrechos utilizados para la navegación internacional en la Convención de naciones Unidas sobre el Derecho del mar. In: Temas de Derecho internacional. Festschrift Frida M. Pfirter de Armas Barea, Buenos Aires, pp 211–226

Carreño Gualde V (1999) La protección internacional del medio marino mediterráneo. Madrid

Charney JI, Alexander L (eds) (1993–1998) International maritime boundaries, 3 vols. Martinus Nijhoff, Dordrecht

Chevalier C (2004) The project of International Marine Park in the Mouths of Boniface in international law. IUCN

Chi Young Pak (1988) The Korean Staits. Martinus Nijhoff, Dordrecht

Churchill RR, Lowe AV (1985) The law of the sea. Manchester University Press, Manchester

Ciciriello MCl (1996) Le droit international de la mer: points et contradictions. In: Ben Achour R, Laghmani S (eds) Harmonies et contradictions en droit international. Ed. A. Pedone, Paris, pp 311–335.

Clingan ThA (1994) The Law of the Sea. Ocean Law and Policy. Austin & Winfield, London

Colombos CJ (1967) The international law of the sea. Longmans, London

Coloquio de Madrid (1983) Coloquio Internacional sobre la factibilidad de una comunicación fija a través del Estrecho de Gibraltar, Madrid 9 al 13 noviembre 1982, 2 vols, SECEG, 1983

Comité V (1997) La incidencia de la ratificación española del Convenio de Montego Bay sobre el régimen jurídico del estrecho de Gibraltar. REDI 1997:322–333.

Couper A (ed) (1983) The Times atlas of the oceans. Times Book, London

Cuyvers L (1986) The Strait of Dover. Martinus Nijhoff, The Hague

Dahak D (1986) Les Etats arabes et le droit de la mer. Tome I, Les Editions Maghrébines

Dean (1958) The Geneva Conference on the Law of the Sea: what was accomplished. Am J Int Law 1958:607 et seq

Deroucher FG (1972) Freedom of passage through international straits: community interest and present controversy. University of Miami Press, Florida

Díez-Hochleitner J (1986) Régimen de navegación de los buques de guerra extranjeros en el mar territorial español y de sus escalas en puertos españoles. REDI 1986.2:543–569

Division de Asuntos Oceanicos y del Derecho del Mar (1995) El derecho del mar. La práctica de los Estados en el momento de la entrada en vigor de la CNUDM. Naciones Unidas, New York

Djamchid M (1974) Du droit de passage dans le détroit de Tiran. Rev Egypt Droit Int 1974: 27–35

Dunlap W (1996) Transit passage in the Russian Arctic Straits. Durham University Press, Durham

Dupuy RJ, Vignes D (1985) Traité du nouveau droit de la mer. Economica, Paris

Dyoulgerov M (1999) Navigating the Bosporus and the Dardanelles: a test for the international community. Int J Marine Coastal Law 1999:57–100

El-Baradei (1982) The Egyptian–Israeli peace treaty and access to the Gulf of Aqaba: a new legal regime. Am J Int Law 76:532–554

Elliot-Meisel E (2009) Politics, pride and precedent: the United States and Canada in the Northwest passage. Ocean Dev Int Law 40.2:204–232

El-Mor (1981) The regime of passage in straits used for international navigations in the light of the Third UN Conference on the Law of the sea. Rev Egyp Droit Int 37:13–83

Espaliú Berdud C, Casado Raigón R (2005) El paso inocente de buques de guerra extranjeros por mar territorial: alcance del régimen de la CNUDM. REDI LVII(2):777–798
Fink JE (1995) The Gulf of Aqaba and the Straits of Tiran: the practice of "freedom of navigation" after the Egyptian–Israeli Peace Treaty. Naval Law Rev 1995:121–144
Fornari M (2005) Conflicting interests in the Turkish Straits: is the free passage of merchant vessels still applicable? Int J Marine Coastal Law 2005:225–246
Frank EJ (1981/1982) UNCLOS III and the straits passage issue: the maritime powers' perspective on transit passage. New York Law School J Int Comp Law 3:243–270
George M (2002) Transit passage and pollution control in straits under the 1982 Law of the Sea Convention. Ocean Develop Int Law 2:189–205
George M (2004) The regulation of maritime traffic in straits used for international navigation. In: Oude Elferink AG, Rothwell DR (eds) Oceans Management in the 21st century: institutional frameworks and responses. Martinus Nijhoff, Leiden, pp 19–48
Gestri (1986) Liberti di navigazione e prevenzione dell'inquinamiento: il caso dello stratto di Messina. Rev Diritto Int 1986:280 et seq
Ghosh (1980) The legal regime of innocent passage through the territorial sea. Ind J Int Law 1980:216
Gidel G (1934) Le droit international public de la mer. Le temps de paix, t III, Paris
Giulano (1975) The regime of straits in general international law. Italian Yearbook of International Law, pp 16–26
González Campos JD (1975) Navegación por el mar territorial incluidos los estrechos. In: Poch A (ed) La actual revisión del Derecho del Mar. Una perspectiva española, I. Primera Parte, Madrid, pp 285–397
Grandison WG, Meyer VJ (1975) International straits, global communications and the evolving law of the sea. Vanderbilt J Transnational Law 8:393–449
Grotius H (1925) Del derecho de la guerra y de la paz (trad J. Torrubiano), Madrid, p. 324
Grotius H (1956) De la libertad de los mares (Trad. V. Blanco García y L. García Arias), Madrid
Grunawalt (1987) United States policy on international straits. Ocean Dev Int Law 18:445–458
Guiliano M (1975) The regime of Straits in General International Law. Italian Yearbook International Law, pp 16–26
Guillaume G (1990) Détroits internationaux et droit aérien. Ann Air Space Law 1990:75–91
Hamzam BA (1996) The Straits of Malacca and the law of the sea. In: Proceedings of the International Law Association (ILA), pp 15–30
Harlow BA (1985) UNCLOS III and conflict management in straits. Ocean Dev Int Law 15: 197–208
Hart HLA (1961) The Concept of Law. Clarendon, Oxford
Hill (1933) Le regimen international des détroits maritimes. R des C 45:479–555
Ho JH (2009) Enhancing safety, security, and environmental protection of the Straits of Malacca and Singapore: the cooperative mechanism. Ocean Dev Int Law 40.2:233–247
Hyde (1945) International Law, Chiefly as Interpreted and Applied by the United States. Little, Brown & Co, Boston
Infante MT (1995) Straits in Latin America. The case of the Strait of Magellan. Ocean Develop Int Law 2:175–187
International Hydrographic Organization (2006) A manual on technical aspects of the United Nations Convention on the Law of the Sea 1982, 4th edn. International Hydrographic Organization, Monaco
Jayewardene HW (1990) The Regime of Islands in international law. Martinus Nijhoff, Dordrecht
Jeanel R (1984) Les grands compromis de la Conférence. In: SFDI, Perspectives du droit de la mer à l'issue de la 3ª Conférence des Nations Unies. Colloque de Rouen. Pedone, Paris
Jia, Bing-Bing (1998) The regime of straits in international law. Clarendon, Oxford
Jiménez Piernas CB (1982) El proceso de formación del Derecho internacional de los archipiélagos, 2 vols. Editorial de la Universidad Complutense, Madrid

Jiménez Piernas C (2001) La ratificación por España de la Convención de 1982 sobre Derecho del Mar y el Acuerdo de 1994 sobre aplicación de la Parte XI: Nuevos riesgos de codificación del Derecho internacional. REDI 2001(1–2):105–124

Johnston DM (2002) The Northwest Passage revisited. Ocean Develop Int Law 2:145–164

Joyner ChC (2002) Regulating navigation through the Turkish Straits: a challenge for modern environmental law. Int J Marine Coast Law 4:521–560

Juste Ruiz J (1990) Los estrechos internacionales en el Derecho del Mar. In: VIII Semana de estudios del mar, Gijón, pp 117–128

Karagiannis S (1999) L'élément fonctionnel dans la définition des détroits internationaux. In: Mélanges offerts à Emmanuel Langavant, pp 169–220

Kaye St. () The Torres Strait Treaty: a decade in perspective. Int J Marine Coast Law 1994 (3):311–336

Kaye StB (1997) The Torres Strait. Martinus Nijhoff, Dordrecht

Kempton StB (2000) Ship routing measures in international straits. Ocean Yearbook, pp 232–250

Knight HG (1972) The 1971 United States proposals on the breadth of the territorial sea and passage through international straits. Oregon Law Rev 1972:759–788

Koh KL (1982) Straits in international navigation: contemporary issues. Oceana, New York

Krafft M (2009) The Northwest Passage: analysis of the legal status and implications of its potential use. J Marit Law Comm 40.4:537–578

Labrousse (1984) Les détroits dans la Convention. In: SFDI, Perspectivas de droit de la me à l'issue de la 3ème Conférence de Nations Unies, Paris, pp 100–116

Lagoni R (2003) Environmental protection and maritime safety in the Danish Straits. In: Jochen Abr. Frowein (ed) Liber Amicorum Tono Eitel, pp 600–625

Langdon JBRL (1990) The extent of transit passage. Some practical anomalies. Marine Policy 1990:130

Lapidoth R (1972) Les détroits en droit international. Pedone, Paris

Lapidoth R (1983) The Strait of Tiran, the Gulf of Aqaba and the 1979 Treaty of Peace between Egypt and Israel. Am J Int Law 77:84–108

Lapidoth-Eschelbacher R (1982) The Red Sea and the Gulf of Aden. Martinus Njhoff, The Hague

Larson DL (1979) Security, disarmament and the law of the sea. Marine Policy 1:40–58

Larson DL (1987) Innocent, transit and archipelagic sea lanes passage. Ocean Develop Int Law 18(4):411–444

Lay SH, Churchill R, Nordquist M (1973) New directions in the law of the sea. Documents, vol 2. Oceana, London

Lawyer JE (1974) International straits and the law of the sea conference. Air University Review, sept–oct 1974

Leanza U (1999) Il Diritto degli spazi internazionale. Parte Prima. La Tradizione, Torino

Leifer M (1978) Malacca, Singapore and Indonesia. Sijthoff and Noordhoff, Alphen aan den Rijn Sijthoff

Levy LP (1983) La Conférence des Nations Unies sur le droit de la mer: histoire d'une négociation singulière, Paris

López Martín AG (2002) Tratados sucesivos en conflicto: criterios de aplicación, Servicio de Publicaciones de la Facultad de Derecho de la UCM, Madrid

López Martín AG (2006) Conflicto entre tratados. Tempestad a calma en el Derecho del Mar? Foro: Nueva Época 3:241–278

Lucchini L, Voelckel R (1990) Droit de la Mer, Tome 1. Pedone, Paris (Tome 2, 1996)

MacDonald CD, Mitsuyasu CA (2000) Regulatory setting for very large floating platforms in Hawaii. Ocean Coastal Manage 43:73 et seq

Macfie AL (1993) The straits question 1908–1936. Institute for Balkan Studies, Thessaloniki

Macnab R (2004) Canada's Arctic waterways: future shipping crossroads? Meridian 2004:1–6

Maduro MF (1980) Passage through international straits: the prospects emerging from the third United Nations Conference on the Law of the Sea. J Marit Law Comm 12(1):65–95

angone GJ (1987) Straits used for international navigation. Ocean Develop Int Law 18:391–409

Mankabady S (1986) The International Maritime Organization, vol 1, international shipping rules. Croom Helm, London

Marffy A de () La Déclaration Pardo et le Comité des Fonds Marins. In: Dupuy RJ, Vignes D (1985) Traité du Noveau Droit de la Mer. Ecomica, Brussels, pp 123–142

Marín López A (1979–1981) Los estrechos internacionales y la III Conferencia de Naciones Unidas sobre Derecho del Mar. ADI 5:46–64

Marín López A (1984) El paso de los estrechos. Cursos de Derecho Internacional de Vitoria-Gasteiz, pp 63–98

Mayama A (1995) The influence of the straits transit regime on the law of neutrality at sea. Ocean Develop Int Law 1995:1–30

Merani, Sterling (1971) Legal consideration of the Israeli–Egyptian dispute involving the right of innocent passage through the Straits of Tiran. Ind J Int Law 1971:411 an seq

Meseguer Sánchez JL (1999) Los espacios marítimos en el nuevo derecho del mar. Marcial Pons, Madrid

Momtaz D (1974) La question des détroits à la troisième conférence des Nations Unies sur le droit de la mer. Ann Français Droit Int 1974:841–859

Moore JN (1980) The regime of straits and the third United Nations conference on the law of the sea. Am J Int Law 1980:77–121

Muench (1981) Reflections on the emerging straits and the draft convention on the law of the sea. Ind J Int Law 1981:31–327

Nandan SN (1982) The provisions on straits used for international navigation in the United Nations Convention on the Law of the Sea. Singapore J Int Comp Law 1998:393–396

Nandan SN, Anderson DH (1989) Straits used for international navigation, a commentary on Part III of the United Nations Convention on the Law of the Sea of 1982. BYBIL, pp 159–204

Nandan SN, Rosenne S (eds) (1993) United Nations Convention on the Law of the Sea 1982: a commentary, vol II. Martinus Nijhoff, Dordrecht

Ngantcha F (1990) The Right of Innocent Passage and the Evolution of the International Law of the Sea. The Current regime of 'free' navigation in coastal waters of third States. Pinter, London

Nolta (1974) Passage through international straits: free or innocent? The Interest at State. San Diego Law Rev 1974:815 et seq

Nordquist MH et al (1985) United Nations Convention on the Law of the Sea, a Commentary. Martinus Nijhoff, Dordrecht

Norsquist MH (ed) (1993) United Nations Convention on the Law of the Sea 1982. A commentary, vol II. Martinus Nijhoff, Dordrecht

O'connell DP (1975) The influence of Law on Sea Power. Manchester University Press, Manchester

O'connell DP (1982) The International Law of the Sea, vol 1. Clarendon, London (vol 2, 1984)

O'neil WA (1998) The IMO and international straits. Singapore J Int Comp Law 1998:284–289

Office for Ocean Affairs and the Law of the Sea (1989) The law of the sea. Baselines: an examination of the relevant provisions of the United Nations Convention on the Law of the Sea. United Nations, New York

Office for Ocean Affairs and the Law of the Sea (1992) The law of the sea. Straits used for international navigation. Legislative history of part III of the United Nations Convention on the Law of the Sea, 2 vols. United Nations, New York

Oude Elferink AG (2000) The regime of passage through the Danish Straits. Int J Marine Coastal Law 2000:555–566

Oxman BH (1982) Le régime de navires de guerre dans le cadre de la CNUDM. AFDI 1982:811–850

Oxman BH (1985) La troisième Conférence des Nations Unies sur le Droit de la Mer. In: Dupuy RJ, Vignes D (eds) Traité du Nouveau Droit de la Mer. Economica, Bruxelles, pp 143–216

Oxman BH (1995) Environmental protection in archipelagic waters and international straits. The role of the International Maritime Organization. J Marine Coastal Law 10:467–489

Oxman BH (1998) Observations on the interpretation and application of the article 43 of UNCLOS with particular reference to the straits of Malacca and Singapore. Singapore J Int Comp Law 1998:408–426

Pastor Ridruejo JA (1983) La Convención de 1982 sobre Derecho del Mar y los intereses de España. Cursos de Derecho Internacional de Vitoria-Gasteiz 1983:69–103

Pharand D (1977) International straits. Thesaurus Acroasium 7:p 62 et seq

Pharand D, Legault LH (1984) The Northwest Passage: Artic Straits. Martinus Nijhoff, Dordrecht

Piaskowski-Rafowicz S (2004) Les ponts sur les détroits. Annuaire du Droit de la Mer, tome IX, pp 319–345

Plant G (1985) International Traffic Separation Schemes in the New Law of the Sea. Marine Policy 1985:134–147

Platzöder R (1978) Meerengen. ZaoRV 38:710–743

Platzöder R (1983) Third United Nations Conference on the Law of the Sea: Documents, vol IV. Oceana, Dobbs Ferry, NY

Poch A (1975) La actual revisión del derecho del mar: una perspectiva española, I. Primera Parte, Madrid

Pratt LJ (1990) The physical oceanography of sea straits. Kluwer, Dordrecht

Przetacznik (1977) Freedom of Navigation through territorial sea and international straits. Revue de Droit International de Sciences Diplomatiques et Politiques 1977:222 et seq

Pueyo Losa J (1981) El archipiélago oceánico. Regulación jurídico-marítima internacional. International Law Association, sección española, Madrid

Pufendorf S (1712) Le droit de la nature et des gens, on systeme general de principes les plus impotants de la morale, de la jurisprudence et de la politique (trad. J. Barbeirac), t I. Amsterdam

Ramazani RK (1979) The Persian Gulf and the Strait of Hormuz. Sijthoff and Noordhoff, Alphen aan den Rijn Sijthoff

Rao (1984) Superpowers strategic interests in straits: a missed bargaining lever of the Group of 77. Ind J Int Law 24:398–405

Reisman M (1980) The regime of straits and national security: an appraisal of international law making. Am J Int Law 1980:48–74

Remiro Brotóns A (1975) Archipiélagos e Islas. In: Poch A (ed) La actual revisión del Derecho del Mar. Una perspectiva española, I. Primera Parte, Madrid, pp 433–530

Rigaldies F (1987) Le statut des eaux de l'archipiel de l'Arctique canadien. Espaces et Ressources Maritimes 2:47–105

Riphagen W (1980) La navigation dans le nouveau droit de la mer. RGDIP 84:144–177

Riphagen W (1984) Le droit de passage inoffensif après la 3º Convention. In: SFDI, Perspectivas de droit de la mer à l'issue de la 3ème Conférence de Nations Unies, Paris, pp 190–209

Riquelme Cortado R (1990) España ante la Convención sobre el derecho del mar. Las declaraciones formuladas, Murcia

Roach JA, Smith RW (1996) United States responses to excessive maritime claims, 2nd edn. Martinus Nijhoff, The Hague

Roberts J (2006) Compulsory pilotage in international straits: the Torres Straits PSSP proposal. Ocean Dev Int Law 2006:93–112

Robertson HB (1980) Passage through international straits: a right preserved in the third United Nations Conference in the Law of the Sea. Vanderbilt J Int Law 5:801–857

Rothwell DR (1992) International straits and UNCLOS: an Australian case study. J Marit Law Comm 1992:461–483

Rothwell D, Sam Bateman (eds) (2000) Navigational Rights and Freedoms and the New Law of the Sea. Kluwer, The Hague

Rowny K (1987) The right of passage through straits used for international navigation and the United Convention on the Law of the Sea. Polish Yearbook of International Law, pp 57–84

Sánchez Rodríguez LI (1981) Estatuto jurídico de los estrechos y Derecho del mar. El proyecto de enlace fijo entre Europa y África por el Estrecho de Gibraltar. Anuario de Derecho Marítimo 3:73–120

Schachte W, Bernhardt PA (1993) International straits and navigational freedoms. Virginia J Int Law 1993:527–556

Schwarzenberger VG (1957) International law as applied by international courts and tribunals, vol I, 1st Part, 3rd edn. Stevens and Sons, London

Scovazzi T (1995) Management regimes and responsibility for international straits with special reference to the Mediterranean straits. Marine Policy 2:137–152

Scovazzi T (2001) The baseline of the territorial sea: the practice of the Arctic states. In: Oude Elferink AG, Rothwell DR (eds) The law of the sea and polar maritime delimitation and jurisdiction. Kluwer, The Hague, pp 69–84

Scovazzi T, Francalanci G, Romano D, Mongardini S (1989) Atlas of the straight baselines, 2nd edn. Dott. A. Guifre, Milan

Shyam M (1975) International straits and ocean law. Ind J Int Law 1975:18

Singapore Conference on the Straits of Malacca And Singapore (1998) Singapore J Int Comp Law

Smith BD (1988) State responsibility and the marine environmental: the rules of decision. Clarendon, Oxford

Solodovnikoff P (1980) La navigation maritime dans la doctrine et la pratique soviétiques. Etude de droit international public, Paris

Spadi F (2001) The bridge on the Strait of Messina: "lowering" the right of innocent passage? Int Comp Law Q 2:411–419

Stelakatos-Loverdos MC (1998) The contribution of channels to the definition of straits used for international navigation. Int J Marine Coastal Law 1998:71–89

Stevenson, Oxman (1975) The third United Nations conference on the law of the sea. The 1974 Caracas session. Am J Int Law 1975:1–30

Suárez de Vivero JL (1985) El nuevo orden oceánico. Consecuencias territoriales, Sevilla

Talaie F (1998) The current legal status of transit passage: analysis of state practice before and after the LOSC. The Indonesian Q 26:52–65

Tillman P (1994) Strait of Malacca and the Law of the Sea. Am J Int Law 1994:885–898

Treves T (1985) La navigation. In: Dupuy RJ, Vignes D (eds) Traité du Nouveau Droit de la Mer. Economica, Bruxelles, pp 687–810

Treves T (1986) Notes on transit passage through straits and customary law. In: Bos A, Siblesz H (eds) Realism in law-making. Martinus Nijhoff, Dordrecht, pp 247–259

Treves T (1990) Codification du droit international et pratique des Etats dans le droit du le mer. R des C 223(IV):9–302

Treves T (ed) (1997) The law of the sea. The European Union and its member states. Martinus Nijhoff, The Hague

Truver SC (1980) The Strait of Gibraltar and the Mediterranean. Sijthoff and Noordhoff, Alphen aan den Rijn Sijthoff

Tymchenko L (2001) The Northern Sea route: Russian management and jurisdiction over navigation in Arctic seas. In: Oude Elferink AG, Rothwell DR (eds) The Law of the Sea and Polar Maritime Delimitation and Jurisdiction. Kluwer, The Hague, pp 269–291

United Nations (1991) UNCLOS 1982: a commentary. Martinus Nijhoff, Dordrecht

Unlu N (2002) The legal regime of the Turkish Straits. Martinus Nijhoff, The Hague

Valle Gálvez, A., El Houdaígui R (2006) Las dimensiones internacionales del Estrecho de Gibraltar, Dykinson

Van der Mensbrugghe Y (2001) Lés détroits Méditerranée. In: Revue de l'Indemer. Les zones maritimes en Méditerranée. Actes du colloque, Monaco, 4–6 Octobre 2001, Pedone, p 55 et seq

Vandyke JM (1996) Legal and practical problems governing international straits. Ocean Yearbook, pp 109–125

Vandyke JM (2009) Transit passage through international straits. In: Chircop A, McDorman TL, Rolston SJ (eds) The future of the ocean regime building: essays in tribute to Douglas M. Johnston. Martinus Nijhoff, Leiden, pp 177–232

Vattel E de (1834) El Dereho de gentes o los principios de la ley natural aplicados a la conducta y a los negocios de las naciones y de los soberanos (trad. M.M. Pascual), t I, Madrid

Visscher Ch de (1969) Problèmes de confins en droit international public, Paris
Vries Lenths P de (1983) The right of overflight over straits states and archipelagic states: developments and prospects. Netherlands Yearbook Int Law 14:165–225
Wainwright RA (1986) Navigation through three Straits in the Middle East: effects on the United States of being a non-party to the 1982 Convention on the Law of the Sea. Case Western J Int Law 18:361–414
Weller M (1985) Straits used for international navigation, national security and customary international law: the US decision not to sign UNCLOS III. Thesaurus Acroasium 15:867–879
Wengler W (1974) La crise de l'unité de l'ordre juridique international. In: Mélanges offers à Ch. Rousseau, Paris, pp 329–340
Westlake J (1904) International Law, vol I. London
Yakemtchouk R (1987) Le Mediterranen orientale dans la politique de puissances détroits
Yakemtchouk R (2003) Le régime de la navigation à travers le détroit de Baltiisk. Revue du marche commun et de l'Únion Européenne, pp 225–230
Yturriaga Barberan JA (1981) Estatuto jurídico del estrecho de Gibraltar y consecuencias de la construcción de una obra fija. Anuario IHLADI 6:185–217
Yturriaga Barberán JA (1991a) Straits used for international navigation. A Spanish perspective. Martinus Nijhoff, Dordrecht
Yturriaga Barberan JA (1991b) El sobrevuelo por los estrechos utilizados para la navegación internacional. Anuario IHLADI 9:359–421
Yturriaga Barberán JA (1993) Ámbitos de soberanía en la CNUDM. Una perspectiva española, Madrid
Zeiden (1979) The emergence of the Gulf of Aqaba problem. Rev Égypt Droit Int 1979:1 et seq

FSC
www.fsc.org
MIX
Papier | Fördert gute Waldnutzung
FSC® C083411

Kolibri 360 GmbH
Ferdinand-Jühlke-Straße 7
99095 Erfurt, Deutschland
produktsicherheit@kolibri360.de